'R RETURN

6'

EFFECTS OF DEFECTS IN COMPOSITE MATERIALS

A symposium
sponsored by ASTM
Committees D-30 on
High Modulus Fibers and
Their Composites
and E-9 on Fatigue
San Francisco, Calif., 13–14 Dec. 1982

ASTM SPECIAL TECHNICAL PUBLICATION 836
Dick J. Wilkins, General Dynamics,
symposium chairman

ASTM Publication Code Number (PCN)
04-836000-33

 1916 Race Street, Philadelphia, Pa. 19103

Library of Congress Cataloging in Publication Data

Effects of defects in composite materials.

(ASTM special technical publication ; 836)
Includes bibliographies and index.
"ASTM publication code number (PCN) 04-836000-33."
1. Composite materials—Defects—Congresses.
I. Symposium on Effects of Defects in Composite
Materials (1982 : San Francisco, Calif.) II. ASTM
Committee D-30 on High Modulus Fibers and Their Composites.
III. American Society for Testing and Materials.
Committee E-9 on Fatigue. IV. Series.
TA418.9.C6E37 1984   620.1′18   83-73441
ISBN 0-8031-0218-6

NOTE

The Society is not responsible, as a body,
for the statements and opinions
advanced in this publication.

Printed in Ann Arbor, Mich.
September 1984

Foreword

The symposium on Effects of Defects in Composite Materials was held in San Francisco, California, 13–14 December 1982. ASTM Committees D-30 on High Modulus Fibers and Their Composites and E-9 on Fatigue sponsored the symposium. Dick J. Wilkins, General Dynamics, presided as symposium chairman.

Related
ASTM Publications

Long-Term Behavior of Composites, STP 813 (1983), 04-813000-33

Composite Materials: Quality Assurance and Processing, STP 797 (1983), 04-797000-36

Composite Materials: Testing and Design (6th Conference), STP 787 (1982), 04-787000-33

Damage in Composite Materials, STP 775 (1982), 04-775000-30

A Note of Appreciation
to Reviewers

The quality of the papers that appear in this publication reflects not only the obvious efforts of the authors but also the unheralded, though essential, work of the reviewers. On behalf of ASTM we acknowledge with appreciation their dedication to high professional standards and their sacrifice of time and effort.

ASTM Committee on Publications

ASTM Editorial Staff

Janet R. Schroeder
Kathleen A. Greene
Rosemary Horstman
Helen M. Hoersch
Helen P. Mahy
Allan S. Kleinberg
Susan L. Gebremedhin

Contents

Introduction

The objective of the Symposium on Effects of Defects in Composite Materials was to provide a forum for presentations and discussions on the effects of defects on strength, stiffness, stability, and service life. Defects were considered either to originate from the manufacturing process (such as voids, inclusions, and porosity) or to result from service usage including low-energy impact, ballistic damage, ply cracking, and delamination. Contributions were specifically sought on:

1. Observation and measurement of defect location and size.
2. Experimental evidence of consequences of defects.
3. Analytical models for predicting defect behavior.
4. Observations of failure surfaces influenced by defects.

The underlying motivation for selection of this topic for a symposium and publication was an increasing awareness of the importance of defects as they behave as stress concentrators and failure sites in brittle composite materials. The extensive application of such materials in aerospace vehicles and commercial products fostered the need to understand the interrelationships among the manufacturing processes, the inspection techniques, and the in-service performance.

Probably because of various constraints in the industrial community, most of the contributions were from either university or government researchers. Consequently, the viewpoint of the majority of the papers is an attempt to understand and characterize defects, rather than explore their engineering significance.

All but one of the papers is concerned with carbon-epoxy laminates. This amount of emphasis is appropriate because the aerospace industry is so heavily involved with applications of the various commercial forms of carbon-epoxy.

Most of the papers contribute new experimental observations of the effects of various defects. Several papers concentrated on the careful observation and documentation of failure surfaces influenced by defects. The interactions between ply cracks and delaminations have been especially well-documented.

Some intriguing new methods of analysis are proposed by a number of the papers. These new analyses, coupled with the improved understanding provided by the experimental observations, will add to our ability to evaluate the sensitivity of structures to defects.

The contributions provided to this volume by the authors, the reviewers, and the ASTM staff are gratefully acknowledged.

Dick J. Wilkins
Engineering staff specialist, General Dynamics,
Fort Worth, Texas; symposium chairman.

Surendra K. Joneja[1] and Golam M. Newaz[1]

Fracture Toughness and Impact Characteristics of a Hybrid System: Glass-Fiber/Sand/Polyester

REFERENCE: Joneja, S. K. and Newaz, G. M., **"Fracture Toughness and Impact Characteristics of a Hybrid System: Glass-Fiber/Sand/Polyester,"** *Effects of Defects in Composite Materials, ASTM STP 836,* American Society for Testing and Materials, 1984, pp. 3–20.

ABSTRACT: In order to understand the damage mechanism in a glass-fiber/sand/polyester hybrid composite, it is essential to study the effects of inherent flaws or defects on the damage growth in the material. The irregular shape and presence of sharp geometric corners in the sand particles, voids, and improper interfacial bonding are factors that contribute to the weakening of the composite performance. One of the parameters influencing the defect formation is size of sand particles.

In this investigation, the thickness of glass/polyester layer is varied, while the sand/polyester layer is kept at a constant thickness. Laminates are made using different sand particle dimensions in order to investigate their influence on the performance of the hybrid composite. The combined effect of the defects is quantified by measuring the residual backing toughness provided by the glass/polyester layer after the full crack growth in the sand layer. The laminates having fine-sand particles provide better toughness properties in comparison to the coarse-sand laminates.

Impact studies are performed to evaluate the influence of defects on the hybrid composite behavior when subjected to impulsive loading. The load is applied to the glass/polyester face. The effect of thickness of the glass/polyester layer on damage initiation and propagation due to the impacting tup has been studied. It has been found that the thickness of the glass/polyester layer has a predominant influence on damage growth and mode of failure.

KEY WORDS: composite materials, fatigue (materials), fracture mechanics, chopped strand mat, polyester concrete, glass-fiber/sand/polyester hybrid composite, voids, interface, sand particle size, fracture toughness, backing toughness, impact, total energy, initiation energy, ductility index

The relatively low tensile strength and fracture energy of the polyester/sand composite has been the driving force behind the development of glass-fiber-reinforced polyester concrete [1,2].[2] Properties of the glass-reinforced polyester

[1] Advanced engineers, Owens-Corning Fiberglas Corporation, Technical Center, Granville, Ohio 43023.
[2] The italic numbers in brackets refer to the list of references appended to this paper.

sand composite depend on characteristics of the fibers, the resin matrix, and the fiber/matrix interface.) For better design of the composite, investigators active in the field are engaged in establishing the relationship between the micro and macro behaviors of the composite. Micromechanical approaches to complex materials such as plain and fiber-reinforced concrete are commonly based on multi- (or two-) phase models. Stroeven [3] modeled the hybrid composites based on deterministic as well as probabilistic principles to derive the constitutive relationships. Using this model, he evaluated the stress transfer capability of a cracked region in plain and fiber-reinforced concrete.

Excessive voids and poor interfacial bond adhesion between sand and the matrix are common factors that affect the performance of the material. The irregular shape and size of the sand particles further causes high stress concentration at the interface of the matrix and sand [4]. These defects are potential failure initiators. The microscopic defects may combine together to produce degradation of the sand/fiber/polyester hybrid composites. The presence of defects influences critical load for crack initiation and velocity of crack propagation in the material, thus affecting the performance of crack arresting material such as glass fibers.

Equally critical in the design of polyester concrete are the dynamic properties. Many investigators [5,6] have observed improvements in the impact resistance of cement when glass fibers or some toughening agents are introduced into the system. However, very little published work is available on impact characteristics of polyester concrete. Basic understanding of the behavior of polyester concrete at high strain rates caused by impact may provide insight for a more rational design analysis of the system under dynamic loads.

In this study, the combined influence of voids, sand particle size, and interfacial bond adhesion on fracture toughness and impact behavior of glass-fiber/sand/polyester hybrid composites has been investigated. The thickness of the glass/polyester backing layer is varied, keeping the layer of sand/polyester concrete at constant thickness to study the effect of backing toughness. Average sand particle dimensions are changed in order to understand their influence on the performance of the hybrid composite. The effect of the thickness of the glass/polyester layer and particle size on damage initiation and propagation during impact has been analyzed.

Material and Specimen Preparation

The material used in this investigation is a composite made of E-glass chopped strand mat, M721 (ARATON®), and polyester resin E-737, both manufactured by Owens-Corning Fiberglas Corporation. The M721 is constructed from chopped fine strands randomly oriented and bonded in mat form by a small quantity of high solubility polyester resin. The mat weighs 0.457 kg/m^2 (4.48 N/m^2). The E-737 is an unpromoted isophthalic polyester resin having 3.8 to 4.5% ultimate elongation. For fabrication of the hybrid laminates, a mold made of high-density polypropylene was used. First, the resin was spread on the surface of the mold for uniform wetting and then five layers of the mat were placed one by one, pouring resin on the top of each layer. A roller was employed to squeeze out and enhance the impregnation of the resin. A mixture of 85/15 sand/polyester resin by weight was prepared and poured on the top of the mat polyester layers in the mold to make 12.7-mm-thick laminates. Figure 1 shows a schematic of the laminate in the mold. Sand with two different particle sizes, 100 and 700 μm, were used to make the laminates. The laminates were cured under uniform pressure of 9.65 KN/m² at 22°C for 18 h and postcured at 93°C for 2h.

For fracture toughness tests, single-edge notch beam (SENB) specimens were prepared with three different thicknesses of backing layers, namely, 3.2, 1.6, and 0.8 mm. A notch of 2.5 mm was machined at the center, across the width of the specimen in the sand/polyester layer. A diagram of a finished specimen is shown in Fig. 2a. For the impact study, rectangular cross-section specimens of 12.7-mm width and 127.0-mm length with varying thickness of glass fiber mat polyester layer and constant layer of 9.4-mm sand/polyester were prepared (Fig. 2b). Most of the impact study was performed with samples having the larger sand particle size (700 μm). A limited number of samples with finer sand particles were also subjected to impact.

Experimental Procedure

Fracture Toughness

The microscopic examination of the material revealed that the defects are distributed and oriented randomly all over in the sand/polyester layer. Due to this, many of the defects are not wholly contained in the plane perpendicular to the maximum tensile stress, therefore, not all defects are stressed in a typical tensile mode, K_I. However, an overall fracture toughness is obtained using notched bend tests. The specimens have been loaded at the rate of 1.27 mm/min on an Instron unit. The critical load is obtained from a load-deflection curve.

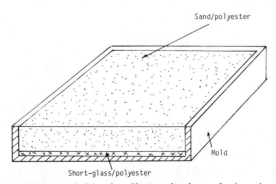

FIG.1—*Schematic of short glass-fiber/sand/polyester laminate in mold.*

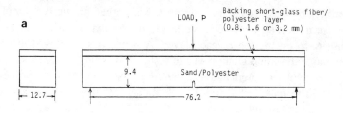

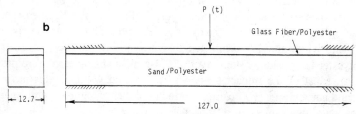

FIG. 2—(a) *Single-edge notched beam specimen for fracture toughness test, and (b)* specimen *for the impact test (all dimensions are in millimetres).*

The fracture toughness is calculated using the following form of the Griffith relationship [7].

$$K_{Ic} = \frac{6M_c a^{1/2}}{b\ w^2}\ Y \qquad (1)$$

where M_c is the applied critical bending moment; a, b, and w are notch depth, width, and thickness of the specimen, respectively; and Y is a dimensionless parameter that depends on the ratio, a/w, and is given by

$$Y = 1.93 - 3.07\ (a/w) + 14.53\ (a/w)^2$$
$$- 25.11\ (a/w)^3 + 25.80\ (a/w)^4 \qquad (2)$$

In the hybrid, the value of M_c is calculated based on the load, P_c, at which crack propagation initiates in the sand layer. This load corresponds to first peak in the load-deflection diagram. The crack initiates at critical load and grows in the polyester concrete and finally hits the fiber-glass-polyester layer that arrests the crack. The residual toughness has been calculated as the area under the load deflection curve beyond full crack growth in the polyester concrete (Fig. 3).

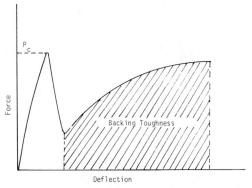

FIG. 3—*A typical load-deflection curve for the hybrid.*

Impact

The Rheometrics High Rate Impact Tester is used to study the impact response of glass-fiber/sand/polyester laminates. The specimens were mounted against a 76.2 mm opening, keeping both ends fixed. A hemispherical tup of 12.7-mm radius was used to impact the specimens on the backing side made of glass-fiber-polyester layer. The impact speeds of 22, 88, and 220 cm/s were selected to determine the influence of velocity of impact on the behavior of the composite. Plots of load-deflection and energy were obtained for different thicknesses of the backing layers.

Some specimens were also subjected to a bumping type impact. The ram displacement was controlled in order to simulate a bumping type impact. The depth of penetration was accomplished by using the "Return Point Select" mode on the Rheometrics High Rate Impact Tester, thus allowing for only partial sample deformation or surface fracture. To do this, the desired penetration depth is entered into the computer memory, then when the ram advances to the pre-selected penetration depth, a "data-stop" sensor is activated. The ram decelerates and returns to its initial position. An overshoot will occur due to the momentum of the ram and deceleration time. The ram velocity was set at 22 cm/s. The actual depth of penetration as well as the amount of surface fracture propagation will reflect the impact resistant characteristics of the material.

Results and Discussion

The optical and scanning electron microscope (SEM) photomicrographs of the composites reveal that the number of voids per square inch in the fine-sand/polyester layer is higher than in the coarse-sand/polyester concrete (Fig. 4a). However, the average ratio of the major lengths across the biggest void in the coarse sand and the biggest void in the fine sand is approximately seven to eight (Fig. 4b). This may be attributed to the difference in total surface areas of fine-

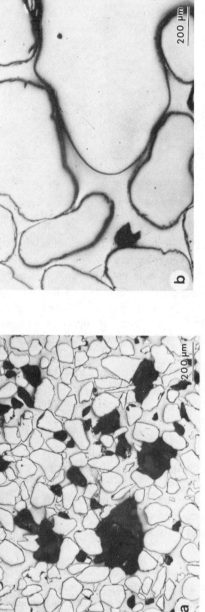

FIG. 4—Photomicrographs showing voids and packing of (a) fine-sand and (b) coarse-sand particles in the polyester cements.

and coarse-sand particles in the laminates. The size of the particles also influences the distribution of the polyester that in turn affects the impregnation and wetting of the sand. The untreated fine and coarse particles provide poor interfacial adhesion to resin as shown in Fig. 5. The combined effects of these defects on the performance of the materials are investigated through fracture toughness and impact tests.

Fracture Toughness

The notch beam test in three-point bending is employed to measure the fracture toughness. At least seven specimens are tested for each of the different composite laminates. The load deflection curves for the polyester concrete without glass fiber are shown in Fig. 6a. Using Eq 1, the value of fracture toughness is calculated based on critical load responsible for crack initiation. The fracture toughness of the fine-sand concrete is about 1.6 times that of the coarse-sand concrete. Past the critical load, slope of the load deflection curve indicates the rate of crack growth. The crack growth in the fine concrete is slower because more energy is consumed to open new free surfaces ahead of the crack tip and the crack path is more tortuous. This is due to the smaller particle size and void in the fine-sand/polyester concrete that in turn leads to improved mechanical properties. The photomicrographs reveal that the crack travels along the inter-facial boundaries of the sand particles and the polyester (Fig. 6b). In the coarse concrete, the particles and large voids are responsible for decrease in stresses in steps beyond the critical load, indicating the velocity of the crack to be discrete.

For the hybrid composite having a different thickness of the backing mat-polyester layer, the critical load increases with an increase in the thickness of the backing layer. At the critical load, for a composite having fine particles, load-deflection response is smoother than in the hybrid with coarse-sand particles (Figs. 7 and 8). Due to the addition of glass-fiber layer as the crack arrester, the fracture toughness of the coarse-polyester concrete improves faster in comparison to the fine-sand material. This is attributed to lower stiffness of the coarse concrete. The combined effect of particle size and voids is that crack growth velocity in fine sand is slower than the crack growth velocity in the coarse sand. The values of fracture toughness of the hybrid composite has been calculated and summarized in Table 1. The residual backing toughness against the crack growth is provided by the glass-fiber layer in the hybrid. This has been determined as the area under the points, A and B, as shown in force-deflection diagrams in Figs. 7 and 8. Points A and B correspond to the load when the load first drops down at full crack growth in polyester concrete and the maximum load carried by the glass/polyester layer, respectively. The backing toughness increases as the thickness of glass-fiber layer increases. However, for the same thicknesses of the backing layer, the retained backing toughness is more in the hybrids having fine-sand particles than the coarse sand. This may be due to lower stress gradient created by slower crack growth in the fine-sand concrete layer. From

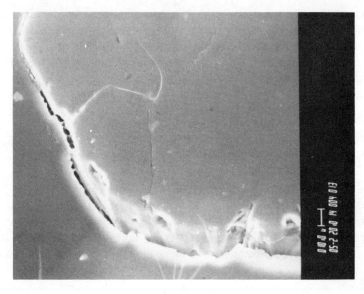

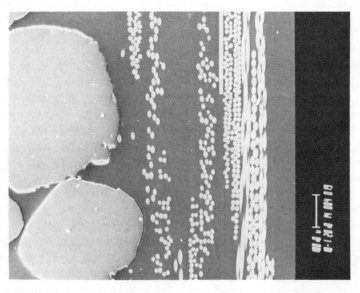

FIG. 5—*SEM photomicrographs showing poor interface between sand particles and resin.*

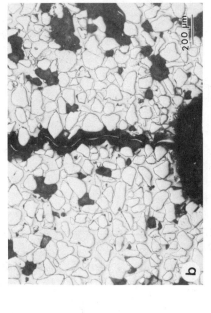

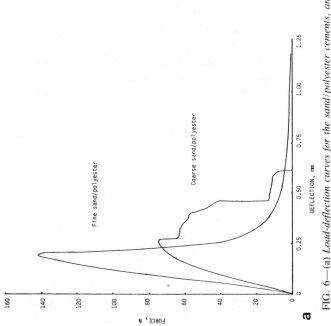

FIG. 6—(a) Load-deflection curves for the sand/polyester cements, and (b) photomicrograph showing path of the crack along the interface between sand and polyester.

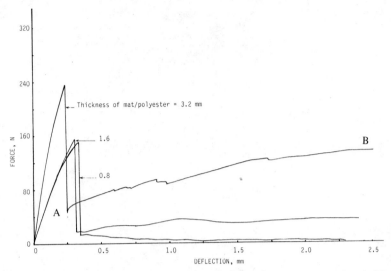

FIG. 7—*Load-deflection curves for the hybrid composites having coarse sand and different thicknesses of the mat polyester (backing layer).*

the cursory analysis of the results, the backing toughness is not a linear function of the glass/polyester thickness. In the coarse-sand hybrid system, 0.8-mm-thick glass/polyester layer does not provide any backing toughness. It may be due to excessive damage in the glass/polyester layer. The selection of proper thickness of the glass-fiber/polyester layer depends on the end-use of the material and its

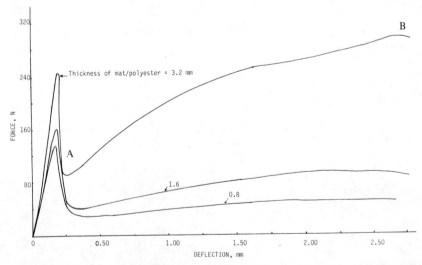

FIG. 8—*Load-deflection curves for the hybrid composites having fine sand and different thicknesses of the mat polyester (backing layer).*

TABLE 1—*Fracture toughness (K_{Ic}) of sand- and glass-fiber/polyester composites having different thicknesses of glass-fiber layers.*

Sand Particle Size, μm	Thickness of Glass/Polyester Layer, mm	Fracture Toughness, $MNm^{-3/2}$	Backing Toughness, Nm
100	0	1.107	0
	0.8	1.221	0.112
	1.6	1.228	0.178
	3.2	1.195	0.304
700	0	0.659	0
	0.8	1.128	0
	1.6	1.166	0.078
	3.2	1.155	0.154

design criterion. Further analysis is needed to determine the backing thickness at which functional damage of the material takes place.

High Rate Impact

A typical impact behavior exhibited by the hybrid samples containing coarse and fine sand are shown in Fig. 9. It is quite clear that both in terms of initiation and propagation energies, the impact resistance of the fine-sand hybrid sample is superior to the coarse-sand hybrid laminate. Impacted specimens demonstrating complete fracture are shown in Fig. 10. The initiation and propagation of these cracks are discussed in the next section.

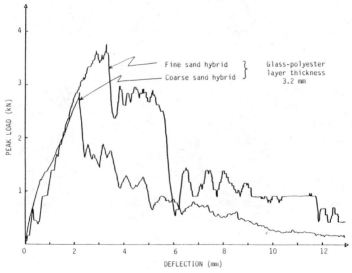

FIG. 9—*Load-deflection curves of fine- and coarse-sand hybrid samples at impact velocity of 88 cm/s.*

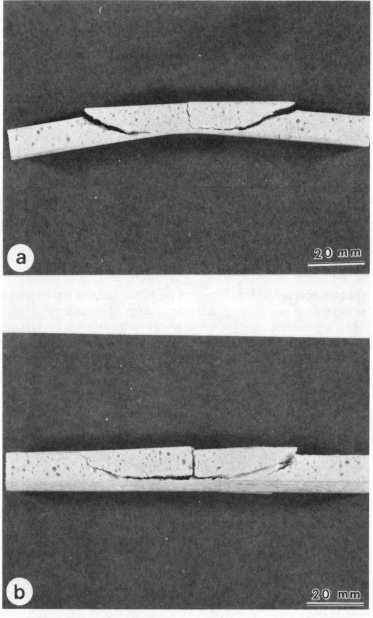

FIG. 10—*Impacted samples showing cracks in the hybrid samples having glass/polyester layer thicknesses of* (a) *0.8 mm and* (b) *3.2 mm.*

One important aspect is to determine the ultimate load carrying capability of the hybrid composites with different glass/polyester layer thickness. For the composites having coarse-sand particles, the ultimate loads carried for impact velocities of 22, 88, and 220 cm/s prior to complete failure is shown in Fig. 11. It is quite clear that the higher thickness of glass/polyester layer has a positive influence on the ultimate load carrying capability. For the constant thickness of glass/polyester layer, the peak load carrying capability does not vary much for different impact velocities between 22 to 220 cm/s. The maximum variation of about 12% is exhibited by the 3.2 mm glass/polyester layer sample. It may be noted that the trend in peak load carrying capability does not change significantly with the increase in backing layer thickness. However, the absolute value of peak load increases as the backing layer thickness increases.

The hemispherical projectile produces time-dependent pressure at the location of impact. Stresses are then generated within the sample. At subsurface locations, triaxial state of stress is produced due to generation of radial, circumferential, and normal stresses. The state of stress in isotropic and composite materials under impact loading is discussed by Greszczuk [8]. For the hybrid composite under investigation, the actual state of stress is not precisely known. To evaluate this, finite element analysis can be used. However, this was not undertaken in this study. Because of the small size of the sample, the impact response is more likely an overall sample response rather than an indication of local deformation.

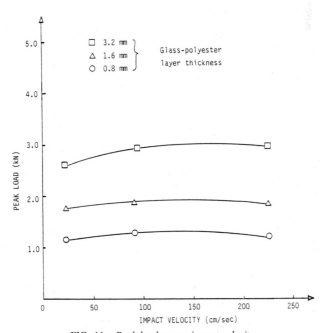

FIG. 11—*Peak load versus impact velocity.*

However, the load-deflection response can be analyzed to distinguish different fracture events within the sample that provides information about the nature of the local deformation. The energy absorbed during impact provides valuable information about the performance characteristics of the material. Two plots (Figs. 12 and 13) are presented showing variation of the initiation and the total energies, respectively, as a function of impact velocity. As shown in Fig. 12, the thicker the glass/polyester layer, the higher the initiation energy. For layer thicknesses of 0.8 and 1.6 mm, the initiation energy either increased slightly or remained constant as the impact velocity was increased from 22 to 220 cm/s. In both these cases, the initiation energy versus impact velocity response is linear. However, the 3.2-mm-layer sample exhibits nonlinear behavior. The energy response of the 3.2-mm-layer sample, between impact velocities of 22 and 88 cm/s increases rapidly and remains about flat thereafter. The early rise of the initiation energy as a function of impact velocity is not well understood for the thicker sample. However, the overall trend of the initiation energy is consistent as is explained later.

The total energy response as a function of impact velocity (Fig. 13) shows that for glass/polyester layer thicknesses of 0.8 and 1.6 mm, the curves pass through maximum energies between impact velocities of 22 and 220 cm/s.

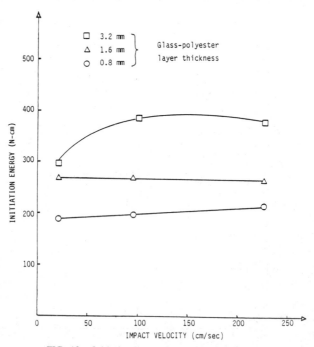

FIG. 12—*Initiation energy versus impact velocity.*

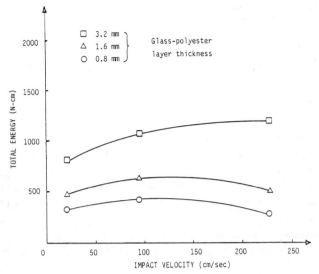

FIG. 13—*Total energy versus impact velocity.*

However, the total energy absorbed by the 3.2-mm-layer sample continuously increases in this velocity range. Thus, the energy absorption capability of the 3.2-mm glass/polyester layer sample is superior to the other samples. Certainly, from an energy absorption viewpoint, it is reasonable to expect that the thicker the glass/polyester layer, the more ductile is the overall composite when subjected to impact loading. The influence of higher backing layer thickness on impact resistance of the hybrid composite under consideration is not well understood. A further investigation on microbehavior of the composite is needed to explain the impact behavior observed in Figs. 11 through 13.

Samples with fine sand having 3.2-mm-thick glass/polyester layer were also subjected to an impact test at 88 cm/s. The peak load-carrying capability is found to increase about 30% in comparison with the coarse-sand hybrid laminate. This trend is also observed for the initiation and the total energies. For the fine-sand hybrid system, both the initiation and the total energies increase about 40 and 50%, respectively. These differences are attributed to the smaller size of the voids in the fine-sand hybrid system. For the coarser sand, interfacial separations, as well as large size of voids make it difficult for a smooth transfer of strain at the glass/polyester and sand/polyester interface. This may result in lower interlaminar shear resistance of the composite. The overall impact resistance of the hybrid laminates then depend significantly on the sand particle size as evidenced here.

By introducing the concept of "Ductility Index" as discussed by Adams [9], the relative degree of brittleness of the composites can be established. The

Ductility Index is defined as

$$D = (E_T - E_i)/E_i \tag{3}$$

where

E_T = total energy, and
E_i = initiation energy.

For the hybrid systems considered, Table 2 shows the values of Ductility Index for various glass/polyester-layer thickness and impact velocities.

As seen in Table 2, the lower the index value, the more brittle is the composite. Also, it is clear from Table 2 that the lower the glass/polyester layer thickness, the more brittle is the hybrid composite at all impact velocities considered. For a fine-sand hybrid system, D is calculated to be 1.86 at impact velocity of 88 cm/s. Comparing to the coarse-sand hybrid system value of 1.5, it is clear that the fine-sand system is more ductile and thus has better energy absorption capability.

Bump Impact

The bump impact tests are performed to determine how the cracks initiate and propagate. Three dominant stages of crack initiation and propagation are identified as illustrated in Fig. 14. Clear cutoff points of delamination and subsequent propagation through the sand/polyester layer are difficult to establish. However, it is observed that delamination occurs prior to oblique crack progression back into the sand/polyester layer.

Summary and Conclusion

The size of sand particles and the formation of void size influence the fracture toughness behavior of polyester concrete. The fracture toughness of the fine-sand concrete is about 1.6 times more than that of the coarse-sand concrete. The size of the particles and voids also affect the crack growth in the concrete.

The critical stress increases with an increase in the thickness of the backing

TABLE 2—*Ductility index of hybrid samples.*

Impact Velocity, cm/s		Ductility Index, D, glass/polyester-layer thickness		
		3.2 mm	1.6 mm	0.8 mm
22	coarse sand	1.8	0.8	0.7
88		1.5	1.1	1.0
220		2.0	1.0	0.4
88	fine sand	1.86	...	...

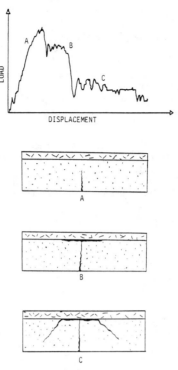

FIG. 14—*Various stages of crack propagation due to impact loading.*

layer. The combined influence of size of the sand particles and voids on the impulsive stresses at the interface between polyester concrete and the glass-fiber/polyester layers is evaluated in terms of residual backing toughness. The backing toughness increases with the increase in the thickness of the glass-fiber/polyester layer. While this conclusion may have been drawn intuitively, the results of this work have quantified its magnitude and offer a method for its measurement. This quantification can aid in designing systems with a desired fracture toughness.

Furthermore, the retained backing toughness is higher in fine-sand hybrid than in coarse-sand hybrid, for the same thickness of the glass-fiber/polyester layer. However, as the backing layer thickness increases, the toughness increase becomes nonlinear.

Overall impact resistance of the hybrid laminates depend significantly on the sand particle size and backing thickness of sand/polyester layer. Comparing to the coarse-sand hybrid system, the fine-sand system is more ductile and has better energy absorption capability.

The present study reveals that improvement in backing toughness and impact resistance of the hybrids having the same thickness of glass-fiber/polyester layer can be achieved by using finer sand.

References

[1] Suaris, W. and Shah, P. S., *Composite,* Vol. 13, No. 2, April 1982, pp. 153–159.
[2] Kobayashi, K. and Cho, R., *Composite,* Vol. 13, No. 2, April 1982, pp. 164–168.
[3] Stroeven, P., *Composite,* Vol. 13, No. 2, April 1982, pp. 129–139.
[4] Durelli, A. J., Parks, V. J., Feng, H. C., and Chaing, F. in *Proceedings,* Fifth Symposium on Naval Structural Mechanics, Pergamon Press, May 1967, pp. 265–336.
[5] Hannant, D. J., *Fibre Cements and Fibre Concrete,* A Wiley Interscience Publication, New York, 1978.
[6] Jamrozy, Z. and Swaney, R. N., *International Journal of Cement Composites 1,* No. 2, July 1979, pp. 65–76.
[7] Griffith, A. A., *Philosophical Transactions,* Royal Society, London, Vol. A221, 1921, p. 163.
[8] Greszczuk, L. B. in *Foreign Object Impact Damage to Composites, ASTM STP 568,* American Society for Testing and Materials, 1975, pp. 183–211.
[9] Adams, D. F. in *Composite Materials: Testing and Design (Fourth Conference), ASTM STP 617,* American Society for Testing and Materials, 1977, pp. 409–426.

Russell D. Jamison,[1] *Karl Schulte,*[2] *Kenneth L. Reifsnider,*[3] *and Wayne W. Stinchcomb*[3]

Characterization and Analysis of Damage Mechanisms in Tension-Tension Fatigue of Graphite/Epoxy Laminates

REFERENCE: Jamison, R. D., Schulte, K., Reifsnider, K. L., and Stinchcomb, W. W., **"Characterization and Analysis of Damage Mechanisms in Tension-Tension Fatigue of Graphite/Epoxy Laminates,"** *Effects of Defects in Composite Materials, ASTM STP 836,* American Society for Testing and Materials, 1984, pp. 21–55.

ABSTRACT: The mechanisms by which subcritical and critical damage develops in several lamination geometries of T300/5208 and T300/914C graphite/epoxy material during tension-tension fatigue were closely examined. A damage analogue in the form of stiffness reduction was used to provide a framework by which the sequence of damage development could be correlated with mechanical response. Stiffness reduction, measured continuously during the course of cyclic loading, was shown to provide a reproducible characteristic correlation with percent of life expended. The relationship was observed to differ markedly among lamination geometries, but for a given geometry was found to clearly indicate the partition of the mechanical response into distinct regions in these characteristic curves. These regions, moreover, were shown to be predominated by particular damage mechanisms—some already discussed in the literature, others less well-recognized.

Results of the observed damage development sequence for cross-ply and quasi-isotropic laminates are presented along with a preliminary association between this damage and the characteristic stiffness reduction curves for these geometries. The geometries used were characterized by distinct, predominant, early subcritical damage conditions. This secondary and subsequent damage development was examined in relation to known, predictable beginning state. Of particular emphasis in each case was the role of this developing damage state in the fracture of fibers in the 0-deg plies.

Damage detection and characterization was accomplished using both nondestructive and microscopic techniques. Two techniques proved to be of considerable utility: penetrant-enhanced stereo X-ray radiography and scanning electron microscopy of coupons taken from penetrant-enhanced deplied, damaged specimens.

A number of significant damage conditions, not heretofore reported, were observed: the production of interior delaminations at the 0/90-deg interfaces of $[0,90_2]_s$ laminates by the gradual growth of longitudinal cracks in the 0-deg plies; the existence of a dense distributed

[1] Assistant professor, Mechanical Engineering Department, U.S. Naval Academy, Annapolis, Md. 21402.
[2] Research scientist, Institut für Werkstoff-Forschung, DFVLR, D-500, Koln 90, West Germany.
[3] Professor, Department of Engineering Science and Mechanics Virginia Polytechnic Institute and State University, Blacksburg, Va. 24061.

microcrack condition at all distinct interfaces of $[0,90,\pm45]_s$ laminates; the segregation of 0-deg fiber breaks in all laminates into zones coincidental with cracks in the adjacent plies; and, the appearance of shear fracture in 0-deg fibers associated with the passage of longitudinal splits.

Mechanisms for each of these damage conditions are proposed in terms of the micromechanics of the predominant damage condition with which they are associated and the global stress state.

KEY WORDS: composite materials, fatigue (materials), damage mechanisms, graphite/epoxy nondestructive evaluation, fibers, X-ray radiography, stiffness changes, delamination, microcracking, matrix cracks, fiber fracture, fracture mechanics

The engineering use of composite materials is rapidly developing both in the sense of the range of different applications and the criticality of the components. For aircraft, for example, complete wing structures and fuselages are being widely planned, and a few are already being flown. These developments bring with them a requirement of reliability during low-level, long-term, variable amplitude cyclic loading of the type that is common to engineering components.

While phenomenological characterization is always required when questions of strength or life must be answered, the ultimate accuracy and success of engineering models of behavior depends greatly on the degree to which they are based on an understanding of the mechanisms that produce damage and reduce the strength and life of engineering components. This is especially true if predictions of behavior are to be made for situations for which no experience is available. It is also important to understand damage mechanisms if improvements in material design and optimization of component design are to be attempted.

The present research effort is concerned with the identification, characterization, and analysis of damage events and mechanisms associated with fatigue loading of graphite/epoxy laminates for a sufficient number of cycles, such that the strength, stiffness, and life of the laminates were reduced. It represents a systematic and comprehensive study of damage development throughout the life of such specimens. Several unique experimental techniques were used to obtain valuable new information about the precise nature of microstructural damage. Several different laminate types and two different materials were used in an effort to identify features that are generic to damage development in laminates under tension-tension cyclic loading. The following sections provide a detailed description of results and a discussion of the implications of these results in the evolution of a general understanding of fatigue damage in composite materials [1].[4]

Experimental Procedure

Laminate specimens measuring 25.4 m wide and 203 mm long were fabricated from T300/5208 (NARMCO) and T300/914C (CIBA-GEIGY) material. The

---

[4] The italic numbers in brackets refer to the list of references appended to this paper.

fabricated ply thickness was approximately 0.14 mm, and the fiber volume fraction was approximately 0.66. Several stacking sequences were chosen to span the range of interlaminar constraint conditions and by anticipation provide a range of damage conditions for study. Results for the stacking sequences, $[0,90_2]_s$, $[0,90,\pm45]_s$, $[0,90,0,90]_{2s}$, and $[0,\pm45,0]_{2s}$, will be reported here. Specimens were subjected to tension-tension fatigue loading at a constant stress ratio, $R = 0.1$, in a sinusoidal form at a cyclic frequency of 10 Hz in a servo-controlled, closed-loop testing machine operating in the load-controlled mode. The maximum cyclic stress amplitude was chosen for each laminate type to produce a lifetime between 10^5 and 10^6 cycles. For example, for the $[0,90_2]_s$ laminate type, this stress was 70% of the static ultimate stress (S_{ult}); for the $[0,90,\pm45]_s$ laminate type, 62% of the static ultimate stress. The ultimate stress in each case was the average of a number of strength measurements for specimens selected from the test population. This choice of maximum stress amplitude produced nearly the same initial maximum 0-deg ply stress in each of the laminate types.

Two series of fatigue tests were conducted for each laminate type. The first was designed to study the sequence of fatigue damage development in a single specimen by nondestructively evaluating its condition at intervals during its fatigue lifetime and will be referred to hereafter as a "stop and go" test. The second series of tests was aimed at producing various levels of expected damage in a number of different specimens, each specimen characterizing a stage of fatigue damage. These specimens were both nondestructively and destructively examined by methods to be described.

Since control of both test series required some inference of the state of damage in the specimen during the course of cyclic loading, a damage analogue in the form of stiffness reduction was employed. The relationship between stiffness reduction and the development of damage in composite material laminates has been studied extensively, and excellent correlations have been reported [2–4]. For the purpose of continuous stiffness monitoring, an extensometer having a nominal gage length of 50.8 mm was attached to the center portion of the specimen. The extensometer knife edges sat in narrow, V-shaped channels machined into metal tabs that were in turn bonded to the specimen surface with silicone rubber cement. The extensometer was held in place with small rubber bands looped around the specimen.

Data acquisition and computation was accomplished by a Z-80 microprocessor-based microcomputer. The calculated quantity taken to represent stiffness was the secant modulus of the dynamic stress-strain curve. It should be noted that laminate stiffness, which is a property of the laminate, and the secant modulus, which is an attribute of the dynamic stress-strain curve, are not one. Stiffness reduction is the more rational damage analogue; secant modulus change is simply a convenient measurable quantity. When static stiffness and dynamic secant modulus values were measured at the same point of a specimen's fatigue life, the dynamic secant modulus was typically higher, the magnitude of the difference depending upon the laminate type. The two quantities were approximately equal for the $[0,90_2]_s$ laminates. However, differences between secant modulus values measured at two points during a fatigue test did not differ significantly from quasi-static stiffness variations measured between the same two points, and the use of former quantity to represent the latter was considered to be justified. The use of stiffness reduction as the damage analogue in the conduct of all fatigue testing was guided by observations made in preliminary tests. In these tests, which were designed to fix the maximum working stress amplitude for each laminate type and for which stiffness was monitored continuously, it was observed that regardless of the stress amplitude and hence the fatigue lifetime, the general form of the relationship between laminate stiffness and percent of fatigue lifetime was unchanged. The form of the relationship was markedly different among the laminate types, but for a given laminate type, exhibited a clear and repeatable structure. Thus, a characteristic curve of stiffness versus cycles could be associated with each laminate type uniquely. Moreover, each of these curves exhibited three distinct regions that provided a framework for the assessment of damage development. In each characteristic curve, the initial state was one of rapid stiffness reduction. This was followed by an intermediate region wherein the stiffness reduction occurs linearly with increasing cycles. The final stage was one of rapid stiffness reduction ending in specimen fracture. These will be designated Stages I, II, and III, respectively, in subsequent discussion of results.

Using these regions of stiffness reduction to establish demarcation points, a stop-and-go series of fatigue tests was conducted for each laminate type. In this series, each specimen was cyclically loaded until the stiffness-versus-cycles curve reached the apparent end of Stage I. The specimen was then removed from the testing machine and examined nondestructively by methods to be described in the following sections. Following this examination, the specimen was returned to the testing machine and data acquisition and stiffness monitoring proceeded until the next selected stiffness reduction level was reached. The examination procedure was then repeated. This stop-and-go process typically continued until the specimen failed in fatigue and intermediate stopping points were chosen frequently enough that each region of damage was included.

The stop-and-go series of tests had the advantage that it provided clear evidence of damage progression in a given specimen. This information was particularly useful in following the development of certain types of damage such as matrix cracking, longitudinal splitting, and delamination. The method had the disadvantage that there were largely uncontrollable extraneous factors involved when testing was interrupted that made quantitative interpretation of the local stiffness reduction and damage development difficult.

For this reason, a complementary series of tests was conducted for each laminate type in which the specimen was cyclically loaded until a desired level of stiffness reduction (and by implication, damage development) had occurred. At that point, the specimen was subjected to first nondestructive and then destructive microscopic analysis but was not subjected to additional fatigue cycles thereafter. This method required that a fairly large number of tests be conducted for each laminate type to provide a collection of damaged specimens that, taken together, were representative of the full range of damage development.

The nondestructive and destructive techniques used for the examination and analysis of damage conditions produced by each of these test series were edge replication, stereo X-ray radiography, and specimen deply. These techniques have been applied successfully by other investigations in damage characterization. However, the present work extended each of the techniques to provide a greater resolution of damage detail than has heretofore been reported. Detailed descriptions of the techniques used are provided in Refs 1, 5, 6, 7, and 8.

Experimental Results

$[0,90_2]_s$ and $[0,90,0,90]_{2s}$ Laminate Types

Figure 1 shows the characteristic stiffness reduction curve for a typical $[0,90_2]_s$ specimen. The $[0,90,0,90]_{2s}$ showed a similar structure with a less pronounced "stair step" character in Stage III. The shape of the curve for short-life and long-life tests is similar and varies little in form from specimen to specimen. Stages I, II, and III are marked on the figure. This partition of the stiffness reduction curve served as well to partition the dominant modes of fatigue damage that occurred in this laminate type, as will be shown in the description of damage for each state of the $[0,90_2]_s$ laminate type.

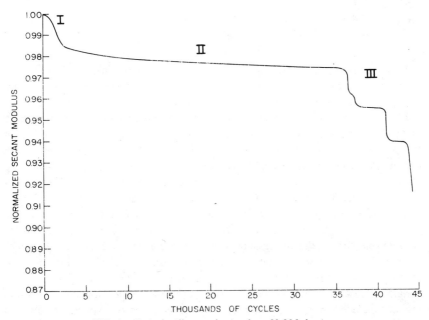

FIG. 1—*Typical stiffness reduction for a $[0,90_2]_s$ laminate.*

Stage I—Figure 2 shows an edge replica of a $[0,90_2]_s$ specimen at the end of 50 000 cycles with a measured stiffness reduction of 2.4% corresponding approximately to the end of Stage I in the characteristic curve. The transverse crack spacing corresponds to the characteristic damage state, a well-established condition of saturated ply cracks [9,10]. Some incipient delamination growth is also observed. Figure 3 is the Stage I portion of the stiffness reduction curve for the same specimen. Also included is a plot of crack density taken from replicas made at intermediate points in Stage I. Approximately one half of the transverse cracks that ultimately form in these laminates do so in the first cycle for the load levels used and cracking is complete at the end of Stage I. The procedure used for starting each fatigue test inevitably resulted in the loss of initial stiffness reduction information. By measuring static stiffness changes at the beginning of a number of tests, it was found that the average stiffness deficit was 4.5%. Thus, the total stiffness reduction in this specimen at the end of Stage I was approximately 6.9%.

Can this stiffness reduction be explained in terms of the formation of transverse cracks alone? The answer is provided by the classical laminated plate theory. The longitudinal stiffness of an undamaged $[0,90_2]_s$ laminate is calculated to be 5.43×10^4 MPa based on the following nominal lamina properties of T300/5208 graphite/epoxy material:

$$E_1 = 14.2 \times 10^4 \text{ MPa}$$
$$E_2 = 10.2 \times 10^3 \text{ MPa}$$
$$\nu_{12} = 0.305$$
$$G_{12} = 6.0 \times 10^3 \text{ MPa}$$

If, for saturation cracking, it is assumed that the longitudinal stiffness E_2 and the shear modulus G_{12} are reduced to zero and if these discounted properties are used in the laminated plate theory, then the predicted laminate stiffness becomes 4.75×10^4 MPa. The laminate stiffness reduction due only to saturation cracking of the 90-deg plies is thus 12.6%. This is more than sufficient to account for the measured stiffness reduction of 6.9%. The fact that it is substantially more can be attributed to the fact that transverse cracking reduces the load-carrying capacity of those plies only in a local region adjacent to the cracks. The material between adjacent cracks outside of these relaxed zones is capable of carrying some load and hence contributing to the laminate stiffness. For example, if saturation cracking is assumed to reduce the longitudinal stiffness of the 90-deg plies to one half of the undegraded value, then the laminate stiffness becomes 5.09×10^4 MPa and the calculated stiffness reduction is 6.3%. Inasmuch as the measured stiffness reduction is acquired over a 50.8 mm gage length and is certainly "global" when compared to the spacing of the approximately 80 cracks that are included therein, the total discount scheme can be expected to provide an upper bound on the actual stiffness reduction.

FIG. 2—*Edge replica from a* [0,90₂]ₛ *laminate at Stage I.*

Other damage at this stage is relatively minor. Some small delaminations confined to a boundary layer along the edge are observed. These delaminations actually appear to mark the beginning of Stage II damage.

Stage II—Figure 4 is a radiograph of a specimen at the end of Stage II. Aside from the transverse cracks, the dominant structures are longitudinal cracks. These cracks are present in Stage I but are few in number and small in length. During Stage II damage development, both measures increase. They exhibit a fatigue character on a macroscopic scale, growing slowly and stably with the increasing cycles. As will be seen from the discussion of Stage III damage development,

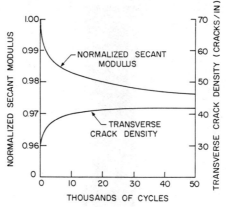

FIG. 3—*Stiffness reduction and crack development for a* $[0,90_2]_s$ *laminate at Stage I.*

this growth is not complete at the end of Stage II. But because the formation of transverse cracks is complete at the beginning of Stage II and other damage modes are only moderately active at this stage, longitudinal cracking predominates.

The key to understanding the formation and growth of longitudinal cracks in 0-deg plies in tensile loading lies in the stress state in that ply. For a uniaxial load in the x-direction, the transverse stress, σ_y, in the y-direction is strongly tensile, owing to the magnitude of the Poisson mismatch between the 0-deg ply and the adjacent 90-deg plies. The transverse strength of the 0-deg plies, however, is low. In fact, at the maximum operating stress levels used for this laminate type, the σ_y stress is approximately equal to the static transverse strength of the 0-deg ply. However, the interlaminar constraint condition prevents complete cracking on the first cycle.

Besides the fatigue aspects of longitudinal crack growth that will be discussed further in the section describing Stage III damage development, a significant and unexpected phenomenon was observed. Figure 5 is an enlargement of a portion of Fig. 4. Of interest are the dark, halolike structures associated with some of the longitudinal cracks. Under stereoptical inspection, each of these structures was seen to be at one of the 0/90-deg interfaces and, by comparison with similar radiographic images at the edges of other laminate types, appeared to be a delamination. Sections of fatigue-damaged specimens in which longitudinal cracks were identified were prepared such that the plane of the cut was normal to the 0-deg fiber direction and placed so as to be adjacent to, but not intersecting, the longitudinal crack. The section surface was then abraded, polished, and inspected microscopically in a repeated cycle until the end of a longitudinal crack was encountered. Figure 6 is a scanning electron microscope (SEM) photograph of the initial encounter of a longitudinal crack in a 0-deg ply. A narrow, irregular crack through the full ply thickness is observed. At the interface, the crack turns downward and travels along the resin-rich zone at the

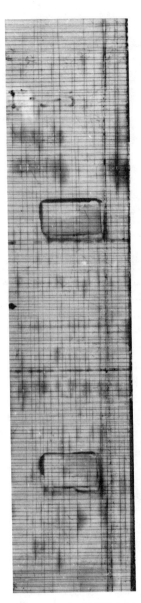

FIG. 4—*X-ray radiograph of a* [0,90$_2$]$_s$ *laminate at Stage II.*

interface. A similar crack turning is frequently observed when cracks in 90-deg plies meet this same interface. Followed to its terminus, this delamination was seen to extend a distance greater than four times the 0-deg ply thickness.

Figure 7 is an SEM photograph of the same crack at a parallel section approximately 0.25 mm from that of Fig. 6. The longitudinal crack is seen to be

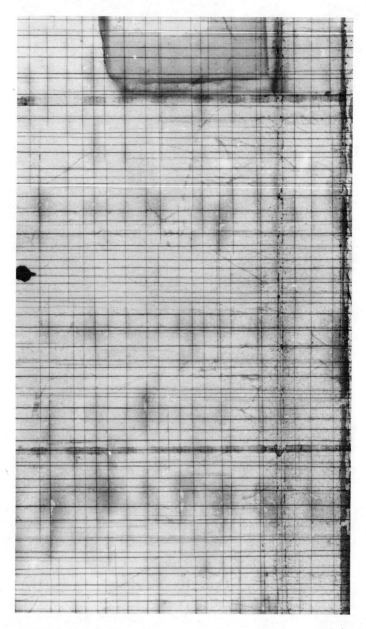

FIG. 5—*Detail of X-ray radiograph of a* $[0,90_2]_s$ *laminate at Stage II.*

FIG. 6—*Transverse section of a longitudinal crack (near tip).*

more widely open, and a second branch of the delamination is evident. The delamination is also wider. Continued section studies of this and other specimens indicate that the opening dimensions of longitudinal cracks and associated de-laminations can be significant when compared to the ply thickness. Successive parallel sections provide a picture of the delamination as a shallow domelike structure with the longitudinal crack as its apex.

In none of the sections examined was a longitudinal crack observed that did not extend completely through the 0-deg ply. This was true of both incipient and well-developed longitudinal cracks. Moreover, no instance was observed when the longitudinal crack was not associated with a delamination. There appears then to be a rapid or instantaneous nucleation step in longitudinal crack development that involves simultaneously the nucleation of a delamination.

Although the *growth* of longitudinal cracks can be attributed to the significant transverse stresses that act on the 0-deg plies of this laminate type, the *nucleation* process is related to the local stress state about the transverse cracks in the adjacent 0-deg plies. Setting aside for the present the anisotropic, inhomogeneous complexities of the problem and treating the transverse crack as a crack in an infinite, homogeneous isotropic plate and assuming plane strain conditions, the stresses in the neighborhood of a crack tip are tensile for a tensile load applied

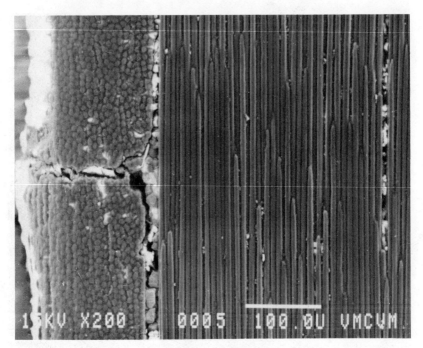

FIG. 7—*Transverse section of a longitudinal crack (away from crack tip).*

perpendicular to the axis of the crack. Nair and Reifsnider [11] have shown by an approximate solution that the stress field about a crack tip in a nonuniform material (with spatially variable mechanical properties) is similar in form to that given by linear elastic fracture mechanics for uniform media, and that for a positive strength gradient at the crack tip, the stresses are tensile. Such a positive gradient exists at the interfaces of fiber-reinforced composite materials.

Figure 8 shows the implications of the existence of a tensile stress, σ_t, in the neighborhood of a crack in the 90-deg plies. The material in the 0-deg ply adjacent to the crack tip is subjected to this tensile stress in its lowest strength direction. Hence, the transverse crack can be considered a likely site for the nucleation of longitudinal cracks. Examination of radiographs in which longitudinal cracks appear indicates that the majority of incipient longitudinal cracks do in fact intersect or are adjacent to the transverse cracks.

The local stress nucleation model can be applied to delaminations as well, as Fig. 9 illustrates. Here the local tensile stress condition is provided by the longitudinal crack. The tensile nature of the stress in this case produces an out-of-plane tensile stress component, σ_t, at the interface. In addition, an out-of-plane tensile stress component, σ_t, is provided by the transverse cracks. At intersections of these cracks, the stresses are additive and the propensity for the nucleation and growth of delaminations should be highest. This is, in fact, observed experimentally. The centers of incipient delaminations frequently co-

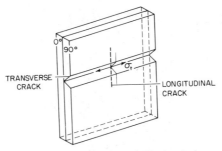

FIG. 8—*Transverse 0-deg stresses in the neighborhood of a transverse crack.*

incide with crack intersections. These are also the points of maximum delamination opening.

Delaminations created in this way grow in size only moderately during Stage II damage development, apparently by coalescence. In the absence of a global driving force, for example, interlaminar shear or normal stress components, they remain a local phenomenon.

Stage III—Figure 10 is a radiograph of a $[0,90_2]_s$ specimen made at the end of 372 000 cycles of loading with a 5.1% stiffness reduction. This stiffness reduction corresponds approximately to Stage III in the characteristic curve. When compared with the Stage II radiograph of Fig. 4, the most notable difference is the increased average length and density of longitudinal cracks. From analysis of radiographs made during stop-and-go fatigue tests, the spacing shown in Fig. 10 is found to represent a saturation longitudinal crack spacing of approximately 1.6 mm corresponding to 0.62 cracks per millimetre. The spacing was determined by measuring distances between adjacent cracks under the stereoptical microscope. The variation in spacing is similar to that observed for transverse cracks. In fact, the rationale for predicting the existence and measure of such a saturation spacing of longitudinal cracks is identical to that applied to the characteristic damage state of off-axis ply cracking. That is, in order for a crack to be formed between two existing cracks, the stress in the ply must attain a value exceeding the transverse ply strength. However, the reintroduction of stress requires a

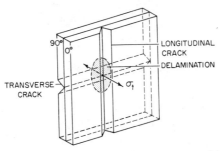

FIG. 9—*Influence of longitudinal and transverse cracks on interior delamination formation.*

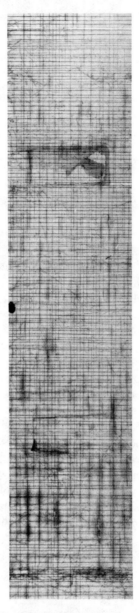

FIG. 10—*X-ray radiograph of a* [0,90₂]ₛ *laminate at Stage III.*

minimum distance in order to be complete. This minimum distance controls the saturation spacing of cracks that corresponds to the characteristic damage state for matrix cracking. By applying a modified one-dimensional shear lag analysis to this laminate type, a longitudinal crack spacing of 1.4 mm corresponding to 0.72 cracks per millimetre is predicted. This compares favorably with the measured value of 0.62 cracks per millimetre. The longitudinal stiffness reduction attributable to a saturation condition of longitudinal cracks is small: 0.5% if total loss of transverse stiffness in the 0-deg plies is presumed in the laminated plate theory; 0.25% if a 50% discount of transverse stiffness in these plies is presumed.

Microscopic examination of Fig. 10 shows that many short, longitudinal cracks exist at this stage. These are cracks that have been nucleated by the local mechanisms described in the previous section and continue to operate in Stage III but cannot grow in the relaxed global stress field created by the growth of their predecessors.

The growth of longitudinal cracks and the coalescence of delaminations associated with them combine to produce a final damage mode characteristic of Stage III development-longitudinal splitting. The condition is shown schematically in Fig. 11. When delaminations coalesce between adjacent longitudinal cracks, a volume of fibers in the 0-deg ply is isolated from the remainder of the material. Over the length of the joined delaminations, fibers in this volume have no load-sharing bond with either the adjacent 0-deg fibers or the 90-deg ply and support the applied load in parallel but independent of the rest of the structure. Fiber fractures in this volume directly reduce the net section and increase the stress concomitantly in the remaining fibers of the volume. As will be shown in a following section, fiber fractures occur fairly frequently in this laminate type. If the accumulation of fiber fractures causes the net section stress to exceed the net section strength anywhere along the section, then failure of all fibers at the section occurs and all load-bearing capability is lost in that volume.

Longitudinal splitting accounts for the "stair step" appearance in Stage III

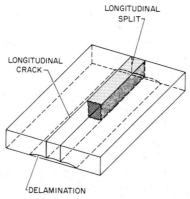

FIG. 11—*Longitudinal split model.*

of the characteristic stiffness reduction curve for the $[0,90_2]_s$ laminate type. At the instant of splitting, the laminate stiffness is reduced by 0.5 to 1.0% through the loss of load-bearing capacity in one of its two load-bearing plies. The effect is diminished somewhat in the $[0,90,0,90]_{2s}$ laminates where the load-bearing contribution of the outer 0-deg plies is proportionally smaller.

These splits frequently occur first at the specimen edge where delamination growth is driven by both longitudinal crack and edge-related mechanisms. Longitudinal splits are observed in the interior of specimens as well. In both cases, the splits often have substantial length. In preliminary tests with this laminate type, the splits were seen to frequently initiate at the gripped ends. However, specimens that had thin epoxy layers applied at the ends were less prone to this behavior. It appears that in the former case, the net section fracture failures that initiated the splits were caused by damage to the 0-deg plies due to gripping.

In summary, fatigue damage development in the $[0,90_2]_s$ and $[0,90,0,90]_{2s}$ laminate types can be characterized as follows: in Stage I the predominant mode is transverse cracking; in Stage II, longitudinal cracks nucleate and grow along the specimen length in the 0-deg plies and produce interior delaminations at the 0/90-deg interfaces; in Stage III these delaminations coalesce in regions between longitudinal cracks, isolating small volumes of material in the 0-deg plies that become longitudinal splits. The measured stiffness reductions can be accounted for by the contributions of the predominant damage modes in each stage of damage.

$[0,90,\pm45]_s$ Laminate Type

The maximum cyclic stress amplitude for all tests was $0.62\ S_{ult}$. Figure 12 shows the stiffness reduction curve for a $[0,90,\pm45]_s$ specimen with Stages I, II, and III marked. The repeatability of this characteristic curve was demonstrated for many specimens representing a range of fatigue lives. It exhibits a markedly different form from that of the $[0,90_2]_s$ and $[0,90,0,90]_{2s}$ laminate types as can be seen by comparison with Fig. 1. The association of damage with the stages of stiffness reduction are provided in the following sections.

Stage I—Replicas and X-rays made at the end of Stage I show that the predominant damage mode is off-axis ply cracking. However, unlike the $[0,90_2]_s$ laminate type, the cracks do not achieve a saturation spacing in any of the off-axis plies at this point although crack formation is well-advanced as Fig. 13 shows. Extension of ply cracking across the width is seen to be incomplete in the $+45$-deg plies and somewhat more complete in the -45-deg and 90-deg plies. In addition, the beginning of edge delaminations at the specimen edges are observed in X-ray radiographs at the end of Stage I.

Stage II—Delamination is the predominant damage mode observed in Stage II for the $[0,90,\pm45]_s$ laminate type as Figs. 14 and 15 show. The stiffness reduction curve associated with this stage is approximately linear. Using a simple analysis proposed by O'Brien, [12] the relationship between delamination growth

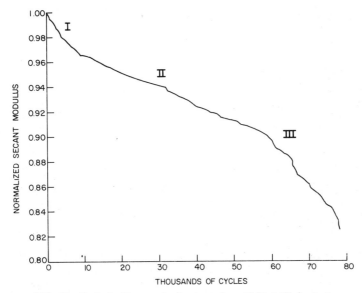

FIG. 12—*Typical stiffness reduction curve for a* $[0,90,\pm45]_s$ *laminate.*

and stiffness reduction for this laminate type was examined in order to quantify the phenomenon.

In principle, the analysis is an extension of the rule of mixtures concept for laminate stiffness. From laminated plate theory, the undamaged laminate stiffness, E_{LAM}, is calculated. Next, a complete delamination in one or more interfaces is assumed. Using the rule of mixtures assumption that the sublaminates formed undergo the same axial strain (but not necessarily the same transverse strain), a fully delaminated stiffness, E^*, is calculated. By assuming that the laminated and delaminated portions of the specimen act as independent components loaded in parallel, the rule of mixtures yields an expression for the partially delaminated laminate stiffness, E, in terms of the ratio of total delaminated area, A, to the total interfacial area, A^*,

$$E = (E^* - E_{LAM})\frac{A}{A^*} + E_{LAM} \qquad (1)$$

Equation 1 then provides a means by which observed delaminations in the $[0,90,\pm45]_s$ laminate type can be related to measured stiffness reduction. For the present case, the quantity E_{LAM} was calculated from laminated plate theory. The quantity A^* was calculated using a sublaminate development based on the experimental observation that in the $[0,90,\pm45]_s$ laminate type, negligible delamination occurred at the 0/90-deg interfaces but substantial delamination occurred at the 90/+45 and +45/−45-deg interfaces. In order to calculate the

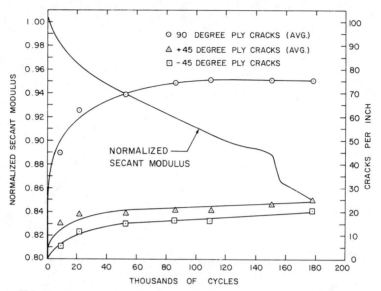

FIG. 13—*Stiffness reduction and crack development in a [0,90,±45]$_s$ laminate.*

delaminated area at each of these interfaces, specimens representing each stage of damage development were treated with gold chloride and deplied. Under proper illumination conditions the gold chloride particles marking the delamination boundaries could be clearly distinguished. By placing a sheet of clear plastic film over the ply, the outline of each delamination could be traced with an ink marking pen. The delamination area was measured with a planimeter. Three measurements of each delamination were made to minimize error and the average value was used. Moreover, since two sets of delamination tracings, one for each facing ply, were made, each delamination area was actually measured twice. The value used for A, the delamination area, was the average of those two measurements. The interfacial area, A^*, for each ply was equal to the specimen width (25.4 mm) multiplied by the extensometer gage length (50.8 mm).

Figure 16 shows the relationship between measured delamination size and measured stiffness reduction due to delamination. The latter quantity represents the difference between the total measured stiffness reduction and the calculated stiffness reduction due to ply cracking. The rule of mixtures prediction from Eq 1 is also shown. Good agreement confirms the linear relationship between stiffness and delamination size for the [0,90,±45]$_s$ laminate type.

Stage III—For the [0,90,±45]$_s$ laminate type, Stage III is a period of accelerating delamination growth, the extent of which is shown in the X-ray radiograph of Fig. 17 where large delamination incursions are observed. It was determined from stereoptical viewing of radiographs that they were located at the 90/+45 and +45/−45-deg interfaces. The net delamination for the specimen shown was approximately 19%.

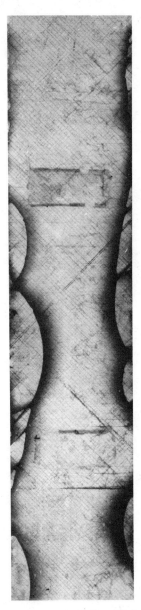

FIG. 14—*X-ray radiograph of a [0,90,±45]ₛ laminate at Stage II.*

Figure 18 is an enlargement of the central portion of Fig. 17. Among the dark lines that are the off-axis cracks can be seen a dense substructure of microcracks aligned with the principal fiber directions. Stereoptical examination of this and other Stage III radiographs place these microcracks at or near ply interfaces. Moreover, the majority of microcracks at a given interface are clustered about

FIG. 15—*Edge replica from a [0,90,±45]$_s$ laminate at Stage II.*

the ply cracks at that interface. The direction of the microcracks appears to be determined by the fiber direction of the ply in which they form. Examples are shown schematically in Fig. 19.

The appearance of microcracks was observed to a lesser extent in other laminate types and the same local tensile crack tip stress field model previously proposed

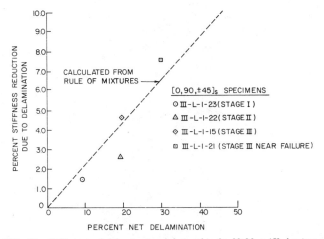

FIG. 16—*Stiffness and delamination relationship for* [0,90,±45]ₛ *laminates.*

can be applied as well to this case. The principal difference among the laminate types is the higher density of microcracks observed in the [0,90,±45]ₛ case. Close examination of radiographs of [0,90,±45]ₛ laminates in Stage II reveals the presence of some microcracks. However, the preponderance of these cracks appears to develop during Stage III. Their role in global stiffness reduction is not likely to be great. Their maximum potential contribution would be to complete the stiffness reduction normally attributed to the saturation condition in the "macrocracks." Their possible effect on fiber fracture is undetermined. If they occur within matrix regions between adjacent fibers in the manner of conventional ply cracks in graphite material, then little effect would be expected. However, if they occur at the fiber/matrix interface and are thereby distinct from the larger ply cracks, then the local effect on fiber fracture might be significant. Continuing work is directed to answering this question.

A rapid, large decrease in stiffness in Stage III was observed in every test. The predominant damage mechanism in Stage III is the continuation of delamination growth from Stage II. Treating the delamination as a planar crack, O'Brien [12] has shown that the concept of strain energy release rate can be applied successfully to describe the delamination growth rate during fatigue cycling. Following his analysis it can be shown that in a constant stress fatigue test of a specimen that exhibits edge delamination, the slope of the stiffness versus cycles curve must increase in magnitude as the maximum strain increases. In the present work, the load-controlled mode employed provided a constant stress condition and the maximum cyclic strain increased monotonically with increasing cycles. Hence, the observed behavior for the stiffness reduction of [0,90,±45]ₛ laminates in Stage III is attributable to the observed rapid growth of delamination in that stage.

In summary, fatigue damage development in the [0,90,±45]ₛ laminate type

FIG. 17—*X-ray radiograph of a* [0,90,±45]$_s$ *laminate at Stage III.*

can be characterized as follows: in Stage I, the predominant mode is off-axis ply cracking; in Stage II, cracking is complete and delaminations grow along the edges and across the width at the 90/+45 and +45/−45-deg interfaces; in Stage III, delamination growth continues until a substantial portion of these interfaces are delaminated at specimen failure.

FIG. 18—*Detail of an X-ray radiograph of a [0,90,±45]ₛ laminate at Stage III.*

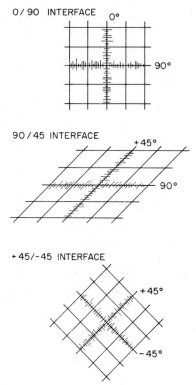

FIG. 19—*Microcracks in* $[0,90,\pm45]_s$ *laminates.*

Fiber Fracture

Discussion of fiber fracture has intentionally been deferred to this point so that the matrix-dominated damage conditions could be developed independently. It would be expected of course that fiber fracture occurs in each of the damage stages for each laminate type. How it develops and what form it takes when it develops will be considered in the next sections.

The scanning electron microscope (SEM) was a principal investigative tool used for this work. Most examinations were performed on coupons cut from deplied laminae. In some cases, portions of edge replicas were also examined in the SEM for evidence of fiber fracture.

Edge Fiber Fracture—Figure 20 is a scanning electron micrograph of an edge replica taken from a fatigue-damaged $[0,90,0,90]_{2s}$ specimen. The cracks in the off-axis plies are evident and individual fibers in each of the plies can be identified. Of interest are the patterns of apparent cracks in the 0-deg plies. At higher magnification (Fig. 21) this crack pattern is seen to be composed of individual fiber fractures in the 0-deg plies. Examination of numerous specimens of this and other laminate types showed that this pattern of fiber breaks was repeated.

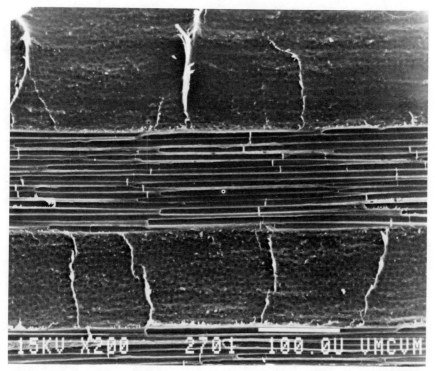

FIG. 20—*Patterns of fiber fractures in an edge replica of a* [0,90,0,90]$_{2s}$ *laminate at Stage III.*

Most significant was the fact that the fiber breaks frequently formed a connected (albeit irregular) path, sometimes extending completely across the 0-deg ply thickness. In addition, the 0-deg ply cracks showed a periodic structure, occurring at repeating intervals of approximately equal distance. Measurements of these intercrack distances in the 0-deg plies confirmed that they were on average equal to the crack spacing in the adjacent off-axis plies. In many cases, the cracks in the two plies were coupled across the interface. This pattern was seen to develop quite early in the fatigue life of these specimens.

Such a crack structure, if repeated across the width of a laminate, has serious implications for laminate failure. In order to examine the phenomenon more closely, the edges of damaged specimens having this structure of fiber fractures were abraded and polished to expose interior fibers for additional replication. The result, in all cases, was that the density of fiber fractures at the interior was found to be greatly diminished. At sections only a few fiber diameters removed from the original specimen edge, there were very few fiber fractures to be seen. This fiber fracture mechanism appears, then, to be a boundary layer effect likely caused by damage to fibers that occurs during specimen fabrication. However, the regular pattern in which the fiber fractures develop and its association with adjacent ply cracking is suggestive of an important link between these damage

FIG. 21—*Detail of fiber fractures in an edge replica.*

modes, which will be discussed in the following section. A more detailed discussion of this portion of the present investigation is provided in Ref *13*.

Interior Fiber Fracture—The 0-deg plies of fatigue-damaged $[0,90_2]_s$ and $[0,90,\pm45]_s$ laminates were examined for evidence of interior fiber fractures. Each specimen had been examined by nondestructive means as described previously. Coupons measuring approximately 6 mm square were cut from each of the deplied load-bearing laminae and examined in the SEM. To assure that the process of deplying did not produce fiber fractures, several virgin specimens were deplied and examined. The fiber arrangement did not show evidence of disturbance either in-plane or out-of-plane as a result of deplying. Moreover, fiber fractures were not observed.

For specimens in Stages I, II, and III of damage development, numerous fiber fractures were observed. These fractures, however, did not occur in a randomly distributed array but instead were observed to have a distinct, consistent pattern, shown schematically in Fig. 22. Zones of fiber breaks occurred within narrow bands that were separated by zones in which there were very few fiber breaks. This pattern was repeated along the length of each specimen. Figure 23a shows a region of fiber breaks, and Fig. 23b is an adjacent fiber fracture-free zone taken from a typical specimen. Coupons taken from near the edge showed the same pattern as coupons from the center of the width.

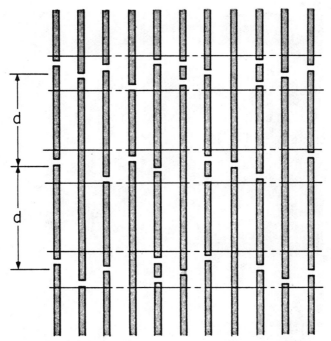

FIG. 22—*Pattern of interior fiber fractures* (d = *adjacent ply crack spacing*).

The pattern of fiber breaks in $[0,90_2]_s$ laminates suggested the involvement of the adjacent transverse cracks whose tips, residing at the 0/90-deg interface, had already been seen to be a fiber fracture initiator at the specimen edge. The question was approached in two ways. First, a series of sequential photographs of coupons in the SEM was taken at ×260 magnification and assembled into lengthwise strips that included a number of fiber fracture zones. The average distance between the centerlines of these zones was calculated for a series of different coupons in the same specimen. This distance was found to be equal to the crack spacing in the adjacent 90-deg plies, approximately 0.64 mm.

Second, several specimens were treated with gold chloride prior to being deplied. The gold chloride entered the transverse cracks in the off-axis plies and left a trace of each crack on the facing 0-deg plies. Under SEM examination, these gold chloride traces were visible, and hence the crack locations were known. For all specimens examined in this way, the zones of fiber breaks coincided precisely with the locations of gold chloride traces. This was true of specimens in each stage of damage and of coupons taken from various locations in those specimens.

As a consequence of these corroborative results and similar results for $[0,90,0,90]_{2s}$ specimens, it was concluded that the preponderance of fiber fractures that occur during cyclic loading of these laminate types result from the influence of transverse cracks at the 0/90-deg interfaces. This mechanism appears

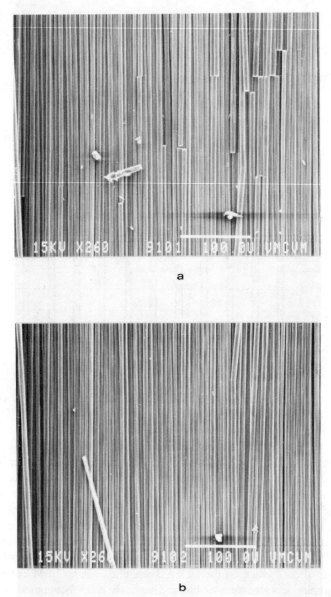

FIG. 23—*Segregation of fiber fractures in the 0-deg ply of a* [0,90₂], *laminate:* (a) *fiber breaks and* (b) *fiber fracture-free zone.*

to supersede any tendency of the fibers to break in a spatially random array determined by the expected strength distribution of single fibers and causes the fiber breaks to occur in a more localized configuration than would otherwise be predicted from statistical strength arguments.

The role of delamination in the fracture of fibers was also examined for the

$[0,90_2]_s$ laminate type. As noted previously, this laminate type delaminates at the 0/90-deg interface across a boundary layer of width approximately equal to the laminate thickness. Coupons were taken from the edge of $[0,90_2]_s$ laminates at the end of Stage I and after edge delaminations had formed. In these coupons, an evident partition between broken and unbroken fibers across each crack trace was observed. Over a distance from the edge approximately equal to the laminate thickness (0.9 mm), few fiber breaks were observed. However, just beyond this point, fiber breaks were frequent and uniformly distributed across the specimen width. This pattern was repeated in other edge coupons.

Several conclusions can be drawn from these results. First, delamination appears to isolate the 0-deg ply fibers from the transverse cracks and thereby prevents related fiber breaks. Second, since crack formation generally precedes delamination in tension-tension cyclic loading of cross-ply laminates, the mechanisms by which transverse cracks produce fiber fractures appears to require a period of time after cracks form before fibers are broken. Third, in light of the previous evidence of fiber fracture at the edge in replicas, it may be necessary to distinguish between mechanisms involving fibers *at* the edge from those involving fibers *near* the edge. The former might be termed a "skin" effect, with the term "edge" effect reserved for the latter.

The role of longitudinal crack growth in the fracture of fibers was also examined for the $[0,90,0,90]_{2s}$ laminate type. Longitudinal cracks are not constrained to grow along only one side of the fiber but can bridge fibers as shown in Fig. 24. As a result of this bridging, a kinking of the fiber occurs locally and produces a shear stress, shown schematically in Fig. 25. The highly anisotropic structure of graphite fibers results in a significant difference between axial and transverse strength, and fibers subjected to combined loading such as this can fail well before the axial failure load in the fiber has been reached. Figure 26 shows an example of a fiber apparently fractured in this way.

The 0-deg and ±45-deg plies of $[0,90,\pm45]_s$ laminates in each stage of damage were also examined for evidence of fiber fracture. Compared to the fiber fracture density observed in the $[0,90_2]_s$ laminates, the $[0,90,\pm45]_s$ fiber fracture density was significantly lower—by an order of magnitude on the average. In fact, there were very few fiber breaks in any ply at any stage of damage for this laminate type. However, the pattern of fiber breaks that were observed were segregated into bands coincidental with the location of cracks in the facing ply just as was seen in the $[0,90_2]_s$ case. This was true not only of fiber breaks in the 0-deg plies caused by transverse cracks in the 90-deg plies but also for fiber fractures in the $+45$-deg plies caused by the action of 90-deg cracks on one face and -45-deg cracks on the other, and similarly for fiber fractures in the -45-deg ply by cracks in the $+45$-deg plies. In each case, the average spacing of fiber fracture zones was the same as the average spacing of ply cracks in the facing plies.

Examination of the mode of individual fiber fractures provides one important distinction between these two laminate types. In the case of the $[0,90_2]_s$ laminates, fiber fracture exhibited a cleavage character with the fracture plane normal to

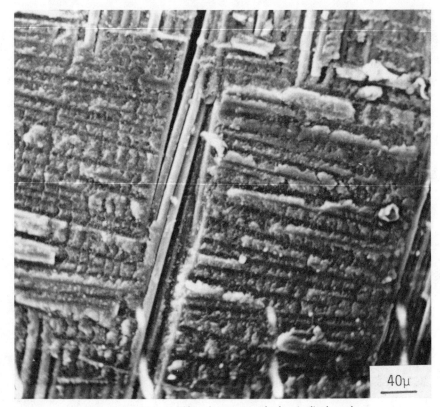

FIG. 24—*Bridging of fibers by passage of a longitudinal crack.*

the fiber axis, as in Fig. 27a. The same appearance was observed for the 0-deg ply fibers in other laminate types, including the $[0,90,\pm45]_s$ type. However, the fracture appearance of fiber breaks in the $+45$-deg and -45-deg plies was quite different. Figure 27b shows an example taken from a $+45$-deg ply. The fiber fractures have a shear mode or mixed mode appearance. With few exceptions, the fiber fractures observed in off-axis plies in $[0,90,\pm45]_s$ laminates at every stage of damage exhibited this apparent shearing influence.

An explanation is provided by considering the state of stress in these plies under tensile loading. In both the $+45$-deg and -45-deg plies, the in-plane shear stresses acting on the fibers are comparable in magnitude to the normal stresses. This situation differs significantly from the stress acting on the 0-deg ply fibers in which the shear component of stress is essentially zero. For the $+45$ and -45-deg plies, the angle of the fiber fracture surface to the fiber axis varied. However, the majority of breaks were observed to be at an angle of approximately 45 deg to the fiber direction.

In summary of the fiber fracture results, the role of off-axis cracks in deter-

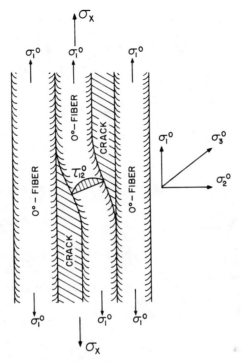

FIG. 25—*Model of fiber kinking by longitudinal crack bridging.*

mining the location and locus of fiber fractures that occur during cyclic loading of graphite/epoxy laminates has been shown. Without exception, the fiber fractures that occur (in large or in small numbers) are segregated into zones that are bounded by fiber fracture-free zones. The stratification of broken fibers in this way produces numerous weakened net sections in each of the load-bearing plies. It also increases the likelihood of the creation of a critical "crack" in the form of multiple adjacent fiber fractures. Whatever the ultimate scenario for the initiation of laminate fracture, the effect of this ordering of fiber fractures cannot be ignored. It is unlikely that dispersed fiber fracture models for unconstrained plies can be successfully used to predict fatigue failure of laminated structures without accounting for the interaction of matrix damage and fiber fracture, possibly as a dominant effect.

Conclusions

While the details of the results of the present investigation are limited to the particular case of tension-tension cyclic loading at elevated stress amplitudes, they are potentially valuable as the basis for suggesting generic aspects of damage development during long-term fatigue loading of composite laminates of the type tested. It is notable that although two different resins were used with the T300

FIG. 26—*Fiber fractured by passage of a longitudinal crack.*

fibers in various laminates, the nature of the microdamage observed was correlated most strongly with the stacking sequence of the laminates examined. Although the resin properties undoubtedly have a role in modifying the details of the damage, the resin constituency was not observed to be a dominant influence.

The chronology of damage development as discussed earlier seems to be a generic aspect of the behavior investigated. Perhaps the most surprising aspect of that chronology is that fiber fracture (if observed at all) occurs early in the fatigue life and no accelerated fiber fracture rate is observed in the late stages when other damage rates do increase (and strength drops sharply). Another generic aspect appears to be the strong relationship of damage modes to one another. Matrix cracks cause local fiber fracture and local interior delamination, especially if matrix cracks in adjacent layers cross each other. Perhaps the most important aspect of this generality is that specific locations where damage initiates tend to become subsequent "centers of local activity" for severe damage concentration. A third generality might be stated as a verification that edges of laminates do experience very special damage development reflecting the special stress states *and* the free surface effect on deformation constraints at that location. Edge behavior requires careful analysis and observations.

FIG. 27—*Fiber fracture modes* (a) *normal and* (b) *shear*.

Finally, several new detailed findings should be noted in closing. The occurrence of interior regions of delamination near matrix cracks is a new damage mode that appears to be very universal in occurrence (at least in fatigue of graphite/epoxy laminates) and a very important fatigue damage development mechanism. It is also surprising to have found that fiber fractures are essentially non-existent in regions away from matrix cracks in adjacent plies, even when large changes in stiffness and strength near the end of fatigue life have occurred. The fact that no evidence was found to support the claim that accelerated fiber fracture may account for the rapid increase in damage rate that is usually observed near the end of life (sometimes referred to as "sudden death") in these laminates was also a salient aspect of the investigation.

Acknowledgments

The authors gratefully acknowledge the financial support of the United States Air Force under Contract Number F33615-81-K-3225, and the guidance, assistance, and encouragement of the contract monitor, Dr. G. P. Sendeckyj (Air Force Wright Aeronautical Labs, Dayton, Ohio). They also acknowledge the support of the Institut für Werkstoff-Forschung, DFVLR (West Germany) that provided financial assistance for Dr. K. Schulte during his visit to the Materials Response Laboratory at Virginia Polytechnic Institute and State University where this work was conducted. We also acknowledge the support provided by the University of Bath (England) for certain activities relating to preparation of the manuscript and to Mrs. Barbara Wengert (Virginia Polytechnic Institute and State University) for typing the manuscript.

References

[1] *Damage in Composite Materials, ASTM STP 775,* K. L. Reifsnider, Ed., American Society for Testing and Materials, 1982.
[2] O'Brien, T. K. and Reifsnider, K. L., "Fatigue Damage: Stiffness Strength Comparisons for Composite Materials," *Journal of Composite Testing and Evaluation,* Vol. 5, No. 5, 1977.
[3] O'Brien, T. K. and Reifsnider, K. L., "Fatigue Damage Evaluation Through Stiffness Measurements in Boron-Epoxy Laminates," *Journal of Composite Materials,* Vol. 15, No. 1, 1981.
[4] Highsmith, A. L. and Reifsnider, K. L., "Stiffness Reduction Mechanisms in Composite Laminates," *Damage in Composite Materials, ASTM STP 775,* American Society for Testing and Materials, 1982.
[5] Stalnaker, D. O. and Stinchcomb, W. W., "Load History-Edge Damage Studies in Two Quasi-Isotropic Graphite Epoxy Laminates," *Composite Materials: Testing and Design (Fifth Conference), ASTM STP 674,* American Society for Testing and Materials, 1979.
[6] Sendeckyj, G. P., Maddux, G. P., and Porter, E., "Damage Documentation in Composites by Stereo Radiography," *Damage in Composite Materials, ASTM STP 775,* American Society for Testing and Materials, 1982.
[7] Rummel, W. D., Tedrow, T., and Bunkerhoff, H. D., "Enhanced Stereoscopic NDE of Composite Materials," AFWAL Technical Report 80-3053, Air Force Wright Aeronautics Laboratory, 1980.
[8] Freeman, S. M., "Characterization of Lamina and Interlaminar Damage in Graphite-Epoxy Laminates by the Deply Technique," *Composite Materials: Testing and Design (6th Conference), ASTM STP 787,* American Society for Testing and Materials, 1982.

[9] Reifsnider, K. L., Henneke, E. G., and Stinchcomb, W. W., "Defect-Property Relationships in Composite Materials," AFML-TR-76-81, Air Force Materials Laboratory, 1979.
[10] Bader, M. G., Bailey, J. E., Parvizi, A., and Curtis, P. T., "The Mechanisms of Initiation and Development of Damage in Multi-Axial Fiber-Reinforced Plastic Laminates," *Mechanical Behavior of Materials: Proceedings of the Third International Conference*, ICM 3, Vol. 3, 1979.
[11] Nair, P. and Reifsnider, K. L., "A Stress Function Formulation and Approximate Solution of the Unsymmetric Deformation Problem for Cracks in Non-Uniform Materials," VPI-E-73-31, Virginia Polytechnic Institute and State University, 1973.
[12] O'Brien, T. K., "Characterization of Delamination Onset and Growth in a Composite Laminate," *Damage in Composite Materials, ASTM STP 775*, American Society for Testing and Materials, 1982.
[13] Schulte, K., Reifsnider, K. L., and Stinchcomb, W. W., "Enstehenund Ausbreiten von Ermüdungsschädigung in CFK," *Proceedings*, 18th AVK Conference, 5–7 Oct. 1982, Freundenstadt, West Germany (in German).

Samuel B. Batdorf[1] and Reza Ghaffarian[1]

Stress Distributions in Damaged Composites

REFERENCE: Batdorf, S. B. and Ghaffarian, R., "**Stress Distributions in Damaged Composites,**" *Effects of Defects in Composite Materials, ASTM STP 836,* American Society for Testing and Materials, 1984, pp. 56–70.

ABSTRACT: Any micromechanical theory for damage accumulation and ultimate failure in composites must depend on a knowledge of the detailed stress distribution in the damaged state. Lack of this knowledge has been a major roadblock to progress in this field. A shear lag approach to this problem has been proposed by Hedgepeth and Van Dyke for unidirectionally reinforced composites, but it is of limited utility because it contains an unknown parameter.

This paper discusses an experimental technique for finding stress distributions in damaged composites with the aid of an electric analogue. In this approach, a current is passed through a model employing conducting rods to represent fibers and a liquid electrolyte to represent the matrix. The potential distribution in the model is analogous to the longitudinal displacement distribution in the composite, and transverse currents are analogous to shear forces. The analogue is employed in the paper to evaluate the unknown parameter in the Hedgepeth and Van Dyke equation.

KEY WORDS: composite materials, fatigue (materials), micromechanics, composite failure, damage accumulation, stress distribution, electric analogue, shear lag, mechanics of failure, unidirectional composites

The practical utility of fibrous composites is largely due to the fact that they generally do not fail when the first fiber ruptures. Instead, they characteristically continue to carry the increasing load for an extended period during which additional damage accumulates, ultimately causing failure. In most high-strength composites, the fibers employed are brittle. As a result, their fracture behavior is characterized by a size effect that can usually be described with reasonable accuracy using Weibull's formula [1].[2]

$$P_s = 1 - P_f = \exp\left[-L \left(\frac{\sigma}{\sigma_0}\right)^m \right] \qquad (1)$$

---

[1] Adjunct professor and post-doctoral scholar, respectively, Materials Science and Engineering Department, University of California, Los Angeles, Calif. 90024.

[2] The italic numbers in brackets refer to the list of references appended to this paper.

where

P_s, P_f = the probability of survival and failure, respectively;
σ = the applied stress;
L = the length of fiber;
σ_0 = the Weibull scale parameter; and
m = the Weibull shape parameter.

While it would of course be highly desirable to be able to track the damage in complex composites such as cross-ply laminates, in point of fact, at the present time we are only just beginning to understand damage accumulation in uniaxially reinforced composites subjected to tension. It has recently been shown [2,3] that the strength of such a composite varies with its size approximately as shown in Fig. 1. A logarithmic plot of the strength of a single fiber obeying Eq 1 versus its length was shown a long time ago [1] to be a straight line of slope $-1/m$. Thus, the first fiber failure in the composite occurs at the stress given by the straight line having that slope in the figure. In the case of a sufficiently small composite, first fiber failure is followed by failure of the composite as a whole. A somewhat larger composite fails only after occurrence of a double break (which we shall call a doublet). A still larger composite becomes unstable and fails only after a triplet is formed. The failure curve is (approximately) a broken line having segments of slope $-1/m$, $-1/2m$, $-1/3m$, etc., corresponding to instability of singlets, doublets, triplets, etc., as shown in Fig. 1. In the case of a composite of the size indicated by the vertical dotted line, the first fiber breaks at the stress labeled σ_1. As the load continues to increase, additional singlets occur at random

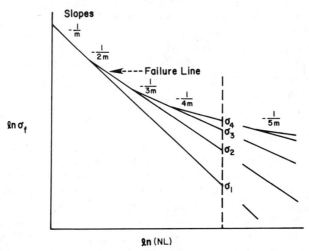

FIG. 1—*Plot of failure stress versus size for unidirectional composite (schematic). Space between top and bottom lines is damage accumulation region.*

locations throughout the composite and, at stress σ_2, one becomes a doublet. With increasing stress, additional singlets and doublets appear, and at stress σ_3 the first triplet appears. Failure, however, does not occur for the case shown in the figure until the first quadruplet appears at stress σ_4. The stress range $(\sigma_1 < \sigma < \sigma_4)$ is the region of damage accumulation. In order to define the failure curve, we need to know not only the slopes of the segments but also the location of the vertices. The latter depend on the values of the Weibull parameters, σ_0 and m, and on the details of the stress distribution in the vicinity of single and multiple breaks (cracks). The stress distribution depends among other things on the size and shape of the crack, on the stiffness ratio of fiber to matrix, on the geometry of the fiber array [for instance whether the construction is two dimensional (single-ply) or three dimensional (multiple-ply)], and on the length of debond, if any, between fiber and matrix.

A number of efforts have been made to establish the stress distribution in damaged composites. Some investigators have employed two-dimensional elasticity theory to describe the matrix behavior and have idealized the fibers as infinitely thin sheets that may either be rigid or elastic [4,5]. Application of the results of such studies is of course limited to the case of single-ply composites.

Dow [6] was among the first to treat the three-dimensional problem. He found the stress distribution in the broken fiber by considering it to be at the center of an elastic cylinder with the properties of the matrix material, and having a diameter such that the volume ratio of fiber to matrix is that of the composite under investigation. While this leads to an estimate of the stress distribution in the broken fibers, the more important question of the stress distribution in the neighboring fibers is left unanswered in this approach.

Using a somewhat more accurate approach, Muki and Sternberg [7] solved the problem of the stress distribution in an infinitely long broken fiber in an infinite elastic matrix. Ford [8] solved the same problem considering both fiber and matrix to be elastic bodies and obtained a surprisingly different result. The two solutions give distances from the location of the break to the location where the fiber returns to 50% of its value before occurrence of the break differing by more than an order of magnitude when the ratio of stiffness of the fiber to stiffness of the matrix is 100.

Shear Lag Solutions

Solutions employing the theory of elasticity are so complicated even for the problem treated by Muki, Sternberg, and Ford that stress distributions in composites must be based on approximate methods of solution. The most commonly used technique is shear lag, which was originally devised to find stress distributions in the sheet stiffener construction commonly used in aircraft. In shear lag theory, the stiffeners (augmented by a certain effective width of sheet) are assumed to carry all the direct stress while the sheet carries only shear. In addition, displacements in both stiffener and sheet are assumed to occur only in the axial direction. Thus, the load per unit length transferred to the ith stiffener by the $(i - 1)$th stiffener is given by the expression

$$\frac{dF_i}{dz} = Gh \frac{(w_i - w_{i-1})}{d} \tag{2}$$

where

F_i = load in the ith stiffener,
z = axial direction,
w = displacement in the axial direction,
h = sheet thickness,
G = shear modulus of the sheet material, and
d = distance between rivet lines.

Equation 2 has been used by a number of authors [9–12] to investigate stress concentration factors or stress distributions or both in two-dimensional composites. The equation should be a good approximation in the two-dimensional case if the sheet of matrix material has a thickness equal to or less than the diameter of the fibers and if the fiber diameter is small compared to the fiber separation. Otherwise, errors of unknown magnitude will be involved, as we shall show after considering the problem of treating a three-dimensional composite in this manner.

Hedgepeth and Van Dyke [13] assumed that the transfer of stress between adjacent fibers in a three-dimensional composite is completely analogous to the case of a two-dimensional composite. Accordingly, for load transfer between two adjacent fibers, they used the equation

$$\frac{dF_{ij}}{dz} = Gh \frac{(w_{ij} - w_{i-1, j})}{d} \tag{3}$$

where the subscripts are defined in Fig. 2. Here, d, the distance between fiber centers, is well defined but it no longer represents the width of matrix material that is deformed in shear. Worse than that, h, representing the sheet thickness in the two-dimensional case, is completely undefined in the three-dimensional case. Also, to always use the sheet thickness for h in the two-dimensional case implies that as $h \to \infty$ the shear coupling between adjacent fibers also $\to \infty$, which is obviously untrue. Without knowing h/d, Eq 3 cannot be used to find the stress distributions in the fibers adjacent to cracks in the composite. Fortunately, this difficulty can be avoided by using an electric analogue.

Electrical Analogue Solution

The basic idea of the electric analogue is illustrated in Fig. 3. The left side of the figure represents a portion of a fibrous composite subjected to tension.

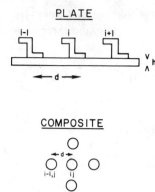

FIG. 2—*Shear lag theory for plates and composites.*

The shear stress is a maximum near the break in the fiber and decreases as the distance from the break increases. The shear stress is proportional to the gradient in the direct stress and therefore to the tensile strain in the composite as shown in the figure. The right-hand side of the figure represents a geometrically similar set-up where the fibers are replaced by conducting rods and the matrix by a fluid electrolyte. A potential applied to the ends of the rod results in a potential gradient

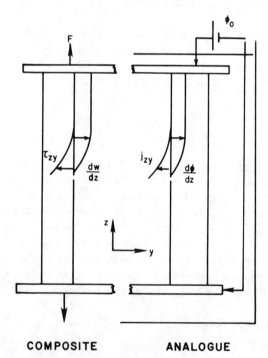

FIG. 3—*Basic concept of electric analogue.*

distribution similar to the strain distribution in the composite, while the transverse current in the electrolyte is similar to the shear stress distribution in the matrix.

In Ref *14*, it is shown that if transverse displacements $u = v = 0$, the equilibrium of a fiber requires that

$$\frac{d^2 w_f}{dx^2} = \frac{G_m}{A_f E_f} \oint \left(\frac{\partial w_m}{\partial y} \, dx + \frac{\partial w_m}{\partial x} \, dy \right) \qquad (4)$$

where the integral is taken around the periphery of the fiber.
Here

    w = longitudinal displacement;
f,m = fiber and matrix, respectively;
  A_f = fiber cross-sectional area;
  E_f = fiber modulus; and
  G_m = matrix shear modulus.

Equilibrium of the matrix in the longitudinal direction requires that

$$\frac{\partial^2 w_m}{\partial x^2} + \frac{\partial^2 w_m}{\partial y^2} + \frac{E'_m}{G_m} \frac{\partial^2 w_m}{\partial z^2} = 0 \qquad (5)$$

Here E'_m is the longitudinal extensional stiffness of the matrix when $u = v = 0$.

The equations for the potential, ϕ, in a geometrically similar electrical set-up are analogous to those preceding except that Eq 5 is replaced by

$$\nabla^2 \phi_e = 0 \qquad (6)$$

They can be made completely analogous by a suitable change in scale. Specifically, if the transverse dimension of the analogue is K times the corresponding transverse dimension of the composite, the longitudinal dimension is chosen to be αK times the composite dimension where

$$\alpha^2 = E'_m / G_m \qquad (7)$$

When this is done, the equations for potential in the analogue are

$$\frac{d^2 \phi_r}{dz^2} = \frac{\rho_r}{\alpha^2 A_r \rho_e} \oint \left(\frac{\partial \phi_e}{\partial y} \, dx + \frac{\partial \phi_e}{\partial x} \, dy \right) \qquad (8)$$

$$\frac{\partial^2 \phi_e}{\partial x^2} + \frac{\partial^2 \phi_e}{\partial y^2} + \alpha^2 \frac{\partial^2 \phi_e}{\partial z^2} = 0 \qquad (9)$$

To complete the analogue, we choose the resistivity ratio to be

$$\rho_r/\rho_e = E'_m/E_r \qquad (10)$$

The analogue solution should be better than the shear lag solution because, in the analogue case, the matrix is assumed to carry its proper share of longitudinal direct stress, and interaction is not limited to nearest neighbors as in shear lag solutions.

The reason for employing such an analogue is that whereas neither the stress nor the strain in the interior of the composite is readily accessible to direct measurement, the potential distribution in the electric analogue can be readily measured using an electric probe. Thus, the interior stress distributions corresponding to simple tension, bending, or shear can be found by measuring the interior potentials when the potentials at the ends of the conducting rods are made to vary in the same manner as the displacements at the fiber ends in the composite vary for these stress states.

The analogue is really quite versatile. The breaks in the fibers are not required to be in the same plane, and, in fact, in naturally occurring cracks they usually are not. Crack interaction can be investigated for either coplanar or noncoplanar cracks. Hybrid composites can be investigated as readily as conventional ones. The effect of frictionless debonding between the broken fiber and the surrounding matrix can be simulated by applying an insulating tape to the appropriate portions of the analogue. Friction between the fiber and the matrix can be simulated by placing a tape with the appropriate resistivity in the appropriate places. Reduction in stiffness due to presence of cracks is easily found. The fractional reduction in stiffness due to the presence of cracks is simply equal to the ratio of the resistivities of the analogue in the undamaged and damaged states.

An electric analogue approach has one limitation in common with numerical methods such as the finite-element approach. That is, a specific solution will apply only for a particular combination of the variables representing the geometry and mechanical properties of the composite under consideration. Therefore, it seems likely that its greatest utility will be to check particular cases, to evaluate the accuracy of other approaches, or to supply missing information that is needed in order to carry out an analytical approach.

Shear Interaction Between Fibers in a Composite

Two-Fiber Interaction

It was brought out earlier that Eq 3 has a very limited utility because no theoretical technique exists for finding h/d at the present time; instead one has to determine this quantity by experiment. The remainder of this paper will be devoted to the problem of how to find h/d. More generally, we want to evaluate the force transfer between fibers in a composite.

Consider a line force on the axis of an infinitely long cylinder of elastic material having a very large radius, R_0. On the surface of a coaxial cylinder of radius r, the shear stress in the axial direction will be

$$\tau = \frac{F'}{2\pi r} \tag{11}$$

where F' is the force per unit length. Since

$$\frac{dw}{dr} = \frac{\tau}{G} = -\frac{F'}{2\pi rG} \tag{12}$$

we can integrate and obtain

$$w = \frac{F'}{2\pi G} \ln r + \text{constant} \tag{13}$$

If we assume that $w(R_0) = 0$, then

$$w(r) = -\frac{F'}{2\pi G} \ln\left(\frac{r}{R_0}\right) \tag{14}$$

We note that since the displacement, w, is constant for all portions of the surface of a cylinder of radius, r, the material inside such a cylinder can be replaced by a rigid rod without changing the external stress field.

If we have two equal and opposite lines of force, the displacement in the elastic body at any point will be the sum of the displacements due to the two force fields. Thus

$$w(x,y) = \frac{F'}{2\pi G} \ln(r_1/r_2) = \frac{F'}{2\pi G} \ln c \tag{15}$$

where $c = r_1/r_2$, r_1 is the distance from the negatively directed line force, and r_2 is the distance from the positively directed line force. Since the locus r_1/r_2 = constant is a cylinder, the preceding equations can be employed to find the shear interaction between any two parallel rods in an infinite elastic medium. This was done in Ref 15, where it was shown that for equal rods

$$h/d = \pi/\ln\left[(1 \pm \sqrt{1 - (1 - s/d)^2})/(1 - s/d)\right] \tag{16}$$
$$= \pi/\ln\left[(1 \pm \sqrt{1 - 4r_0^2/d^2})/(2r_0/d)\right] \tag{17}$$

where s is the distance separating the two rods, d is distance between centers, and r_0 is rod radius. The solid curve in Fig. 4 shows how h/d depends on s/d. Unfortunately, this type of approach cannot be used for more than two loaded circular fibers, because then equal-displacement contours are no longer circular cylinders.

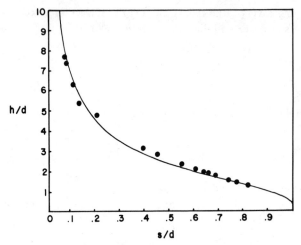

FIG. 4—*Experimental confirmation of theoretical interaction of two infinitely long rigid fibers in infinite elastic matrix.*

The relationship just given for h/d can be checked experimentally using the electric analogue. The rigid cylinders must be modelled by perfectly conducting rods, that is, by rods whose resistivity is many orders of magnitude less than that of the electrolyte. The analogue to Eq 15 is

$$\phi \ (x,y) = \frac{I'\rho_e}{2\pi} \ln c \tag{18}$$

where I' is the current per unit length of conducting rod. From this

$$I' = \frac{\pi(\phi_1 - \phi_2)}{\rho_e \ln c} \tag{19}$$

Thus, $h/d = \tau/\ln c$ varies as

$$I'/(\phi_1 - \phi_2) = R' \tag{20}$$

where R' is the resistance per unit length between the two conducting rods.

The variation in resistance with spacing was measured, and the implied data points for h/d are plotted in Fig. 4. It is seen that the agreement with theory is quite close. Since the theory is exact, discrepancies must be attributed to experimental limitations; such as, inability to obtain an infinitely large electrolytic tank and zero resistance rods, possible contact potential differences between electrode and electrolyte, possible nonuniform conductivity of electrolyte due to ion migration, error in alignment of electrodes and in current measurement, etc.

When only two fibers carry load, but additional fibers are present, the unloaded fibers assume the average displacement the surrounding matrix would have in the absence of such fibers. When fiber radius $r_0 \ll d$, the interaction between the loaded fibers is the same whether the unloaded fibers are present or not. For example if eight coplanar fibers are present, but only the center two are loaded, theory indicates that the displacements in the plane through fiber axes will be as shown in Fig. 5. The expectation from this that current between the electrodes will be unaffected by the presence of additional conducting rods for low fiber volume ratio has been verified experimentally. For high volume ratio, the effective shear stiffness of the medium between loaded fibers (effective conductivity of electrolyte) is increased, so in this case the presence of unloaded fibers does affect the interaction between loaded fibers.

Multiple Fiber Interaction

Consider next the case in which two adjacent fibers in a square array are loaded with forces per unit length $\pm F'$ and the remaining fibers are unloaded. If the fiber volume ratio is low, the displacements in the loaded fibers are given by

$$ w = \pm \frac{F'}{2\pi G} \ln \frac{(d - r_0)}{r_0} \tag{21} $$

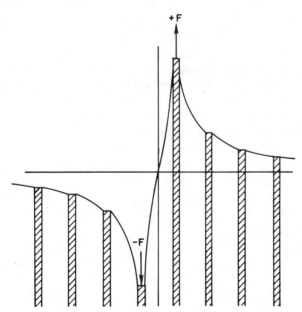

FIG. 5—*Displacements of two loaded and six unloaded fibers in an infinite elastic matrix (schematic).*

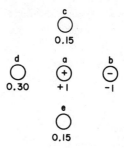

FIG. 6—*Axial displacements of unloaded fibers when loaded fibers are displaced ±1, and when* $d/r_0 = 12.38$.

while the displacements in unloaded fibers are obtained by applying Eq 15. When $d/r_0 = 12.38$ (a value chosen to agree with an experiment performed with the analogue equipment), the theoretical normalized values of w for five fibers, two of which are loaded, are as indicated in Fig. 6. According to Hedgepeth's and Van Dyke's equation, the force on the center fiber depends on its relative displacement with respect to all four surrounding fibers. But, as we have seen, the displacement is the same whether the three unloaded fibers are present or not, which suggests they do not, in fact, interact with the center fiber.

A clue to the resolution of this paradox is found by considering the case of stiffened plates with rigid stiffeners. If only two stiffeners are loaded and they are separated by several unloaded stiffeners, the interaction between the loaded stiffeners is the same whether the unloaded ones are present or not. Nevertheless, a calculation considering them to be present turns out right.

Accordingly, we adopt Hedgepeth and Van Dyke's equation and calculate the force on the center fiber for two different conditions. Under Condition *A*, all outer fibers are displaced an equal distance relative to the center one. Case B will be the situation depicted in Fig. 6.

The idea behind this exercise is that in Case A it is reasonable to assume that the interaction between each peripheral fiber and the center fiber is one quarter of the total interaction. Using this interaction for a fiber pair in a square array composite, we can find its relationship to the interaction between the loaded fibers in Case B, where we know the exact solution (for the limiting case, $r_0 \ll d$) from Ref *15*. This will allow us to evaluate the interaction (value of h/d) for a square array.

In Case A, the Hedgepeth and Van Dyke equation leads to

$$F'_a = 4G \left(\frac{h}{d}\right)_c (w_a - w_b) \qquad (22)$$

where the subscript, c, denotes the value for the two-fiber interaction in a square array composite. In Case B

$$F'_a = G \left(\frac{h}{d}\right)_c [4w_a - w_b - w_c - w_d - w_e] \tag{23}$$

$$= G \left(\frac{h}{d}\right)_i (w_a - w_b) \tag{24}$$

where the subscript, i, denotes the value for a pair of isolated fibers. For sufficiently small r_0/d

$$w_a = -w_b = \frac{F'}{2\pi G} \ln (d/r_0 - 1) \tag{25}$$

Using Eq 15, the displacements of the other fibers are

$$w_c = w_e = \frac{F'}{2\pi G} \ln \sqrt{2} = w_a \frac{\ln \sqrt{2}}{\ln (d/r_0 - 1)} \tag{26}$$

and

$$w_d = \frac{F'}{2\pi G} \ln 2 = w_a \frac{\ln 2}{\ln (d/r_0 - 1)} \tag{27}$$

Thus, for Case B

$$F'_a = G \left(\frac{h}{d}\right)_c \left[4 w_a + w_a + \frac{2 \ln \sqrt{2} \, w_a + \ln 2 \, w_a}{\ln (d/r_0 - 1)}\right] \tag{28}$$

Since $w_a = -w_b$, this can be written as

$$F'_a = G \left(\frac{h}{d}\right)_c (w_a - w_b) [2.5 - \ln 2/\ln (d/r_0 - 1)] \tag{29}$$

$$= G \left(\frac{h}{d}\right)_i (w_a - w_b) \tag{30}$$

Comparing Eqs 29 and 30, we see that

$$\left(\frac{h}{d}\right)_c = \left(\frac{h}{d}\right)_i \bigg/ \left[2.5 - \frac{\ln 2}{\ln (d/r_0 - 1)}\right] \qquad (r_0 \ll d) \tag{31}$$

The relationship between $(h/d)_c$ and $(h/d)_i$ has also been investigated exper-

TABLE 1—*The measured currents.*

					s/d					
	0.84	0.7	0.6	0.54	0.4	0.375	0.3	0.285	0.275	0.2625
I_A	0.186	0.525	0.368	0.361	0.543	0.738	0.980	1.37	2.08	3.2
I_B	0.103	0.279	0.189	0.183	0.269	0.332	0.423	0.548	0.753	1.06
$I_A/4\,I_B$	0.542	0.470	0.487	0.493	0.514	0.556	0.578	0.624	0.6905	0.754

imentally. Five rods were placed in an electrolytic tank in the configuration shown in Fig. 6. A potential of 10 V was applied between Rod a and Rod b. In Case A, Rods c, d, and e were all connected to Rod b by a conducting wire. In Case B, Rods c, d, and e were disconnected and were thus free to adopt the local potential of the electrolyte in which they were immersed. The measured currents were as shown in Table 1, for several different fiber volume ratios.

The last line in the Table gives the ratio of $(h/d)_c$ to $(h/d)_i$. The experimental and theoretical results are compared in Fig. 7. Since Eq 28, representing the theoretical result, is only valid for $r_0 \ll d$ (or $s \rightarrow d$), it is shown as a dotted curve where this condition is not satisfied. The experimental data agree very well with theory in the range where theory is valid, which lends to confidence to their validity. The curve through the data points represents our present best estimate of $(h/d)_c/(h/d)_i$.

The bottom line for this enquiry, of course, is to state what is the absolute value of $(h/d)_c$. Our solution is found by combining the results in Figs. 4 and 7. The result is shown in Fig. 8.

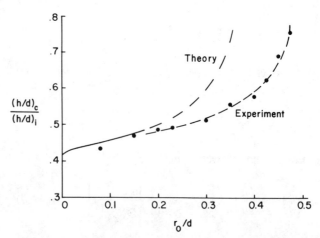

FIG. 7—*Ratio of* h/d *for composite to* h/d *for two isolated fibers.*

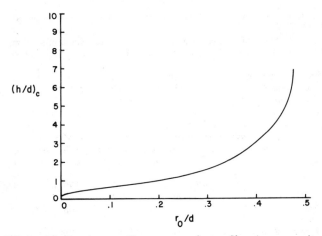

FIG. 8—*Effective shear coupling parameter for two fibers in a composite.*

Conclusion

We have described an electric analogue for finding the local stress distributions in damaged composites. It is closely related to shear lag theory, in that it assumes transverse stresses are zero, but more accurate, since in the analogue the matrix carries its proper share of direct stress, whereas in shear lag the matrix does not. The analogue can also be used to find the reduction in overall composite stiffness resulting from damage and to evaluate certain unknown quantities that must be used in theoretical calculations.

An example of the latter type is the effective shear stiffness of the matrix that appears in the equations of shear lag. This was evaluated first theoretically for very low fiber volume ratio, and then over a wide range of this ratio using the analogue. This result, displayed in Fig. 8, constitutes the first practical application of the analogue and is believed to reduce the uncertainty in the value of effective shear stiffness by more than an order of magnitude.

Acknowledgment

This research was sponsored by the Office of Naval Research under Contract N00014-75-C-0445 and was performed under the cognizance of Dr. Yapa Rajapakse.

References

[1] Weibull, W., "A Statistical Theory for the Strength of Materials," *Ingeniorsvetenskapakademiens*, Handlinger Nr. 151, 1939.
[2] Smith, R. L., *Proceedings Royal Society*, London, Vol. A372, 1980, pp. 539–553.
[3] Batdorf, S. B., *Journal of Reinforced Plastics and Composites*, Vol. 1, 1982, p. 165.
[4] Brussat, T. R. and Westmann, R. A., *International Journal of Solids and Structures*, Vol. 11, 1975, pp. 665–677.

[5] Eringen, A. C. and Kim, B. S., *Letters in Applied and Engineering Sciences*, Vol. 2, 1974, pp. 69–89.
[6] Dow, N. F., "Study of Stresses Near a Discontinuity in a Filament-Reinforced Composite Metal," Space Mechanics Memo #102, General Electric Space Science Laboratories, Jan. 1961.
[7] Muki, R. and Sternberg, E., "Load-Absorption by a Discontinuous Filament in a Fiber-Reinforced Composite," *Journal of Applied Mathematics and Physics (Zeitschrift für Angewandt Mathematik und Physik)*, Vol. 22, Fasc. 5 1971.
[8] Ford, E. F., "Stress Analysis of a Broken Fiber Imbedded in an Elastic Medium," Technical Report No. 1, NSF, GH-33576, June 1973.
[9] Hedgepeth, J. M., "Stress Concentrations in Filamentary Structures," NASA TN D882, National Aeronautics and Space Administration, Langley Research Center 1961.
[10] Ko, W. L., Nagy, A., Francis, P. H., and Lundholm, U. S., *Engineering Fracture Mechanics*, Vol. 8, 1976, pp. 415–422.
[11] Oh, Kong P., *Journal of Composite Materials*, Vol. 13, Oct. 1979, pp. 311–328.
[12] Goree, J. G. and Gross, R. S., *Engineering Fracture Mechanics*, Vol. 13, pp. 563–578.
[13] Hedgepeth, J. M. and Van Dyke, P., *Journal of Composite Materials*, Vol. 1, 1967, pp. 294–309.
[14] Batdorf, S. B., *Journal of Applied Mechanics*, Vol. 50, March 1983, pp. 190–193.
[15] Batdorf, S. B., *Engineering Fracture Mechanics*, Vol. 18, No. 6, 1983, pp. 1207–1210.

Ambur D. Reddy,[1] *Lawrence W. Rehfield,*[1] *and Richard S. Haag*[1]

Influence of Prescribed Delaminations on Stiffness-Controlled Behavior of Composite Laminates

REFERENCE: Reddy, A. D., Rehfield, L. W., and Haag, R. S., "**Influence of Prescribed Delaminations on Stiffness-Controlled Behavior of Composite Laminates,**" *Effects of Defects in Composite Materials, ASTM STP 836,* American Society for Testing and Materials, 1984, pp. 71–83.

ABSTRACT: The effect of delaminations on the stiffness-controlled behavior of laminated composites is investigated. Four panels of $[\pm 45,0,90]_{2s}$ quasi-isotropic layup are fabricated from the T300/Narmco 5208 material system, three with prescribed delaminations of different rectangular shapes and orientations. Nondestructive buckling tests and vibration experiments are conducted on these panels with clamped-free and clamped-simple support boundary conditions to assess the stiffness-controlled behavior. Free-free vibration tests are also conducted to study the influence of delaminations on higher modes. Finite-element analyses are performed to compare with the experimental results. The findings indicate that delaminations of the size considered can be tolerated for some service conditions without dire consequences.

KEY WORDS: composite materials, fatigue (materials), fracture mechanics, composite laminates, composite structures, defects, delaminations, compressive buckling, vibration

The high specific strength and stiffness of composite structures hold great promise for improving the performance of aerospace vehicles. The high potential of graphite/epoxy composites has been demonstrated thoroughly in selected secondary structures, which are generally designed by stiffness requirements. To realize the full benefit of these materials, application in primary structures is currently being emphasized. As primary structures operate at high strain levels and are critical to flight safety, integrity of the material is a paramount consideration.

Unlike metallic materials where failure modes are usually associated with cracking or yielding, laminated composites have several others—delamination being one of the most frequently encountered ones. Delamination may be caused by a manufacturing defect or service-induced. It usually contributes to final failure of the structure. A knowledge of the structural consequences of such

[1] Senior research engineer, professor, and research assistant, respectively, School of Aerospace Engineering, Georgia Institute of Technology, Atlanta, Ga. 30332.

defects on the behavior of composite structures is needed. Stiffness-controlled behavior of graphite/epoxy laminates is investigated in this work. Prescribed delaminations of known shape and size are introduced into the laminated flat panels in order to assess the effect of such flaws on the structural behavior. This effort contributes to the existing data base on this subject [1,2].[2]

In this paper, uniaxial compression buckling and vibration test data on the virgin and delaminated panels are presented. Analytical results obtained with finite-element computer analyses are provided for comparison.

Overview

Bending, buckling, and vibration tests are conducted in the experimental program to determine the relative behavior of the damaged and the undamaged panels. Three- and four-point bend tests are performed to evaluate the panel bending stiffnesses. The panels are then tested in compressive buckling with clamped loaded edges and both free and simply supported unloaded edges. The first test represents a simple situation and the second a more practical one. Buckling tests on the panels are in their fundamental mode with the load estimates obtained from a stiffness plotting technique. As the changes in stiffness should lead to changes in natural frequencies, flexural vibration tests are also conducted on the panels with these boundary conditions. The objective here is to compare the stiffness-controlled behavior from static and dynamic tests for consistency. This may enable use of simpler dynamic tests as our assessment tool in the future work [3,4]. Free-free panel vibration tests are performed to study the influence of imbedded delaminations at higher modes. As the damage affects each mode differently, these data are expected to provide some differences in behavior between panels. Finite-element analyses are carried out by modeling the delamination in two different ways. Structural Analysis Program (SAP) IV is used to estimate the frequencies in both the cases. Buckling analysis, however, is performed only on the second model using the Engineering Analysis Language (EAL).

Experimental Program and Results

Test Specimens

The specimens used in this investigation were fabricated from the unidirectional Thornel 300 graphite/fiber tapes impregnated with Narmco 5208 epoxy resin. Four panels with a $[\pm 45,0,90]_{2s}$ quasi-isotropic layup were manufactured from this material system by Lockheed-Georgia company. One of these has no delaminations and hence serves as a reference or control. Delaminated zones equal to 10% of the panel area, (77.42 cm²), which is thought to be an upper limit to the size encountered in practice between inspections, were built into the

---

[2] The italic numbers in brackets refer to the list of references appended to this paper.

remaining panels between the central plies. These defects were implanted by inserting one ply of TX 1040 fiberglass fabric, manufactured by Palliflex Corporation, sandwiched between two plies of 1-mil (0.00254-cm) thick teflon. The edges of the panels were machined flat and parallel to permit uniform loading. The finished panels measured 25.4-cm wide and 30.48-cm long. They were inspected by ultrasonic C-scan to establish the quality and the size and shape of the imbedded delamination. These C-scan output records are presented in Fig. 1.

The panels were instrumented with strain gages across the width to check uniformity of loading in the compression tests. On the control panel, they were placed close to the center line, whereas, in the case of others, they were located away from the delamination. This is to eliminate possible interaction of local buckling modes with the strain data. The strain gage placement was offset also from the line of inflection determined from the mode shapes generated from a finite-element analysis. In addition, strain gages were mounted at the center and along the edges of delamination. The former is to indicate the local buckling of

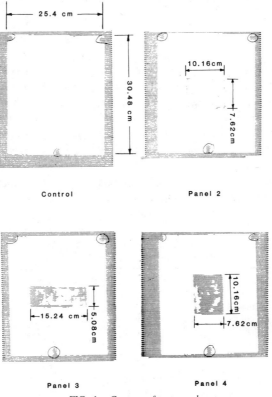

FIG. 1—*C-scans of test panels.*

TABLE 1—*Panel characteristics.*

Panel	Thickness (t), cm (in.)	Weight, kg
Control	0.2057 (0.08097)	0.245
2	0.2099 (0.08264)	0.248
3	0.2092 (0.08235)	0.248
4	0.2076 (0.08173)	0.248

delamination and the latter its propagation during the compression tests. One face of each of the panels was painted with strippable masking film to enable use of moiré-fringe technique to monitor out-of-plane deflections when loaded in compression. The panels were geometrically characterized and weighed before proceeding with the tests. This information is provided in Table 1.

Bend Tests

In order to determine the bending stiffnesses of the panels, three- and four-point bend tests were performed. The load versus central deflection data were generated from the experiments, and the resulting bending stiffness results are presented in Table 2.

Compression Tests

The panels were tested as clamped wide columns under uniaxial compression, a simple situation, to produce buckling in the first mode. Machined angles were clamped to the two opposite edges of the panel to provide the fixed end conditions. Lateral displacements of the panel under increasing load were monitored by a linear variable differential transformer (LVDT) displacement transducer. The buckling load of each panel was estimated from a stiffness plot [5] and is presented in Table 3.

In a more practical configuration, the panels were tested with clamped loaded edges and simply supported unloaded edges. A fixture shown in Fig. 2 was employed to load and support the panel. The load is applied to the specimen

TABLE 2—*Summary of panel stiffnesses.*

Panel	Compressive Stiffness (A), 10^6 N	Bending Stiffness, N · m^2		
		Four-Point Test	Three-Point Test	Computed, $D = \dfrac{At^2}{12}$
Control	26.11	8.12	7.81	9.20
2	28.92	8.23	7.77	10.62
3	27.49	8.41	7.94	10.03
4	26.52	8.49	8.00	9.52

TABLE 3—*Summary of panel buckling test results.*

Panel	Buckling Load (P_{cr}), N		Buckling Coefficient (P_{cr}/Af^2), m^{-2}		Percent Degradation	
	Wide Column	Single Bay	Wide Column	Single Bay	Wide Column	Single Bay
Control	4627	11292	41.882	102.210	...	...
2	4520	11158	35.474	87.572	15.30	14.32
3	4636	11941	38.534	99.253	8.00	2.90
4	4689	12671	41.025	110.862	2.05	-8.46

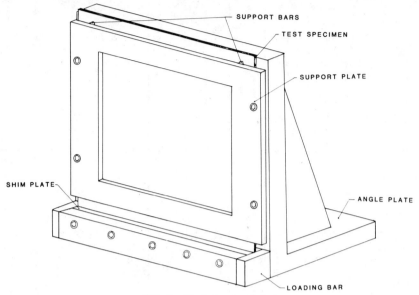

FIG. 2—*Single-bay test fixture.*

through the top loading bar. The shim plates located in the top and bottom loading bars facilitate central location of the panel in the fixture and provide the clamped boundary conditions when tightened. The heavy angle plate is secured to the test machine during testing and has two support bars with rounded edges mounted on its front end. In assembly, the two support bars on the support plate line up with the ones on the angle plate and provide the simply-supported boundary conditions. The fixture is designed to accommodate axial compressive deformation of the panels. Electrical resistance strain gages were used to monitor strain across the panels and around the delaminations. The lateral displacements and end shortening were measured by LVDTs. Acoustic emission instrumentation was also provided on the panels to record the activity during loading. The moiré fringe method was used to check symmetry of support conditions and provide a visual indication of the buckling event. Strain and displacement data were generated on all panels at subcritical loads. Buckling load estimates were obtained using a stiffness plot. They are listed in Table 3. Typical strain data from the back-to-back strain gages on the control panel are presented in Fig. 3.

Compression tests to determine the extensional stiffnesses of the panels were conducted with the same fixture incorporating two extra support bars at one-third panel width. Strain data were recorded from all panels as a function of load and are presented in Fig. 4. Extensional stiffnesses estimated from these data are listed in Table 2.

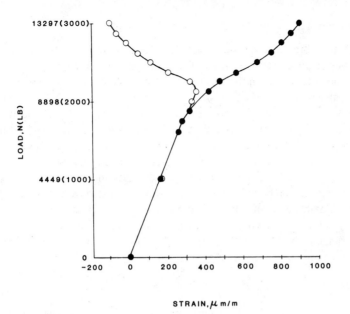

FIG. 3—*Typical control panel back-to-back strain gage response.*

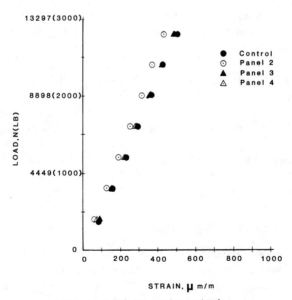

FIG. 4—*Panel compressive strain plots.*

Vibration Tests

In all the vibration tests, the panels were excited by impacting them with a light, instrumented hammer. Dynamic response was sensed by a light-weight accelerometer mounted on the specimen. Vibration data were generated on each of the panels immediately following the buckling tests so that the boundary conditions are preserved. The data on the clamped-free and the clamped-simply supported configurations were analyzed by a Hewlett Packard 5423A Structural Dynamics Analyzer to obtain the natural frequencies, damping, and mode shapes. Frequency information in the two cases is summarized in Table 4.

Free-free vibration tests were performed to study the behavior of the panels at higher modes. The boundary conditions were simulated by hanging the panels by a long, flexible cord attached at a point. This test setup is simple and eliminates boundary influences. Dynamic behavior of the panels was studied in the frequency range 0 to 2400 Hz. This frequency data with the corresponding mode shapes are summarized in Table 5.

Analysis

Finite-element analyses were performed to correlate with the experimental data. In this study, two different models were used to represent the delamination. The first model is shown in Fig. 5 where the panel is divided into two layers of quadrilateral plate elements of equal thickness in the delamination region. They are attached at the nodes by three-dimensional beam elements. Delamination of the required shape can be introduced into the panel by removing the appropriate beam elements. The remaining portion of the panels is made up of plate elements equal in thickness to the panel under study. The bending stiffness of an element with a central plane delamination with continuity condition on lateral displacement and zero shear stress imposed along the delamination length reduces to one-fourth its original value. The second model utilizes these elements, shown in Fig. 6, to represent the delamination zone. Consequently, Young's modulus of these elements is reduced to one fourth in the computations. This is a simpler but weaker model than the first and overestimates the influences. The computer code SAP IV [6] was used to estimate the frequencies of both the models. The data are presented in Table 6. The buckling load information was generated using the Engineering Analysis Language [7] on the second model and the results are summarized in Table 7.

Correlation of Results and Discussion

Bending stiffnesses obtained from the three-point bending tests are lower than those in the four-point tests (Table 2). This is a natural result considering that shear stresses influence the response under three-point loading. The trend of the results is also consistent, and Panel 4 exhibits the highest value for stiffness. The bending stiffness values estimated utilizing the extensional stiffness and the

TABLE 4—*Summary of panel vibration test results.*

Panel	Frequency (ω), Hz		Frequency Coefficient ($\omega^2 m/At^2$), m^{-3}		Percent Degradation	
	Wide Column	Single Bay	Wide Column	Single Bay	Wide Column	Single Bay
Control	158.5	190.00	5.679	8.161	...	...
2	154.6	190.42	4.742	7.194	16.5	11.85
3	156.7	193.5	5.160	7.868	9.14	3.60
4	159.9	192.82	5.655	8.223	0.42	-0.76

TABLE 5—*Free-free panel vibration test data.*

| Mode No. | Frequency, ω(Hz) | | | | Mode Shape |
	Control	Panel 2	Panel 3	Panel 4	
1	102.19	102.14	102.13	102.26	
2	189.21	190.08	190.76	192.24	
3	237.48	236.0	236.21	236.27	
4	270.51	267.3	267.56	268.2	
5	335.23	334.5	335.44	336.88	
6	439.3	438.66	439.4	436.47	
7	481.4	481.0	478.49	482.41	
8	512.7	512.03	514.07	514.14	
9	603.3	599.14	602.53	604.01	
10	646.63	646.68	647.21	645.01	
11	727.96	725.23	728.71	720.45	
12	767.86	766.59	766.32	763.78	
13	853.1	852.4	855	855.6	
	1073	1071	1074	1065	
	1206	1203	1204	1208	
16	1325	1338	1344	1345	
17	1485	1489	1491	1481	
18	1606	1594	1598	1597	
19	1684	1681	1685	1689	
20	1752	1742	1743.7	1747	
21	1967	1968	1972	1961	
22	2055	2050	2056	2044	
23	2144	2140	2144	2154	
24	2228	2224	2223	2223	
25	2388	2384	2395	2389	

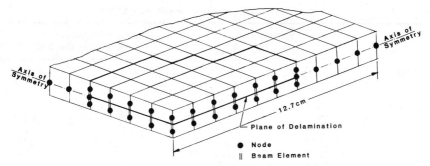

FIG. 5—*Finite-element Model 1 of delamination.*

nominal thickness of panels are listed in Table 2. These are larger than either of the stiffnesses obtained from the bend tests. This simple estimate, which does not account for the individual plies, has been calculated because it utilizes experimentally determined extensional stiffnesses. A reliable means of making a purely theoretical estimate of the bending stiffness for a delaminated panel is not available, although two approaches have been attempted here and are described subsequently.

The buckling load results do not indicate any significant degradation in the stiffness-controlled behavior of the delaminated panels. On the contrary, the load values of Panels 3 and 4 are higher than the control panel. When corrections for variations in panel thickness and extensional stiffness are made, the results look different, however. The percent degradation in both test cases show the same trend, although not in the absolute sense. The dynamic tests on the wide column and single bay configurations also provide similar conclusions when the frequency results are corrected for variations in panel mass, m, extensional stiffness, and thickness.

The analytical frequency and buckling load data presented in Tables 6 and 7 are in a form directly comparable with the results listed in Tables 3 and 4. Model 1 frequency experimental results do not indicate significant differences between panels, whereas large degradations are reflected in the data obtained using Model

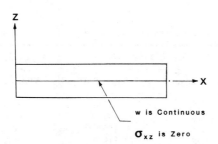

FIG. 6—*Model of a delaminated element.*

TABLE 6—*Summary of frequency results from SAP IV.*

Panel	Frequency (ω), Hz		Percent Degradation Based on Frequency Coefficient	
	Wide Column	Single Bay	Wide Column	Single Bay
		MODEL 1		
Control	216.7	244.9	. . .	. . .
2	226.1	255.9	4.8	4.17
3	221.3	249.5	3.27	3.63
4	217.8	244.8	2.0	2.26
		MODEL 2		
Control	216.9	245.3	. . .	. . .
2	211.2	220.1	16.8	29.3
3	200.4	217.0	20.7	27.3
4	205.1	210.3	12.5	28.1

2. In both cases, the analytical wide column frequencies deviate more from the experimental values suggesting less restraint than the fully clamped condition on the panels. The buckling load results from EAL runs provide satisfactory comparisons with the experimental values. The correlation with single bay test results is good compared to the wide column results. The discrepancy, once again, is due to the boundary conditions and the departure between the analytical and experimental values is of the same order in static and dynamic cases. Model 2 is thus simple, conservative, and could be used to predict behavior of delaminated panels.

The free-free vibration test results in Table 5 include 25 modes of vibration. The absolute differences between the control panel and the panels with delaminations are less than 2% in all the cases considered. The expectation that larger departures in frequency occur at higher wave numbers in defective specimens did not prove so by the data. The imbedded delaminations did not result in changed mode shapes between specimens. Hence, only typical mode shapes are presented in this table.

TABLE 7—*Buckling load predictions by EAL-Model 2.*

Panel	Buckling Load (P_{cr}), N		Buckling Coefficient (P_{cr}/At^2), m^{-2}	
	Wide Column	Single Bay	Wide Column	Single Bay
Control	6726	11554	60.88	104.52
2	7760	13310	60.90	104.46
3	7325	12577	60.88	104.53
4	6958	11945	60.87	104.51

Conclusions

The stiffness-controlled behavior of graphite/epoxy flat panels with prescribed delaminations has been investigated. The data generated are consistent and indicate similar trends in all experiments.

Bending stiffness degradation appears to be negligible when the raw data are examined. The effects are more pronounced after corrections are made for variations in panel thickness, extensional stiffness, and weight. The absolute effects of delaminated zones of the size and shape considered are not large; 16% is the maximum. In all the tests, Panel 4 performed as if perfect in spite of the fact that the delamination is verified by C-scan ultrasonic inspection. This is a significant practical result. It demonstrates that extensive defects can be indicated that have no stiffness-related structural consequence. Furthermore, these delaminations do not influence the panel behavior even at higher modes. These findings suggest that rather large delaminations can be tolerated for some service conditions without serious consequences.

The results of delamination modeling by finite elements, especially with Model 2, provide conservative estimates of behavioral degradation. Further developments in modeling are required to gain better correlation with the experimental data.

Acknowledgments

This work was supported by the United States Air Force Office of Scientific Research under Grant 82-0080. Lockheed-Georgia Company provided the specimens. Their support is gratefully acknowledged.

References

[1] Sprigg, R. G., "An Experimental Study to Determine the Reduction in Ultimate Bending Moment of a Composite Plate Due to an Internal Delamination," M. S. thesis, Navy Postgraduate School, Monterey, Calif., Dec. 1977.
[2] Ramkumar, R. L., Kulkarni, S. V., and Pipes, R. B. in *Proceedings,* 34th Annual Technical Conference, The Society of Plastics Industry, Inc., Section 22-E, 1979, pp. 1–5.
[3] Adams, R. C., Walton, D., Flitcroft, J. E., and Short, D. in *Composite Reliability, ASTM STP 580,* American Society for Testing and Materials, 1975, pp. 159–175.
[4] Kulkarni, S. V. and Frederick D., *Journal of Composite Materials,* Vol. 5, Jan. 1971, pp. 112–119.
[5] Horton, W. H., Cundari, F. L., and Johnson, R. W., "Applicability of Southwell Plot to the Interpretation of Test Data Obtained from Stability Studies of Elastic Column and Plate Structures," USAAVLABS Technical Report 69–32, U.S. Army Aviation Laboratories, Fort Eustis, Va., Nov. 1971.
[6] Bathe, K. J., Wilson, E. L., and Peterson, F. E., "Structural Analysis Program for Static and Dynamic Response of Linear Systems," Report No. EERC 73–11, University of California, Berkeley, Calif., June 1973 (revised April 1974).
[7] Whetstone, W. D., "EISI-EAL: Engineering Analysis Language," Engineering Information Systems, Inc., Saratoga, Calif., Jan. 1979.

Francois X. de Charentenay,[1] Jean M. Harry,[1] Yves J. Prel,[1] and Malk L. Benzeggagh[1]

Characterizing the Effect of Delamination Defect by Mode I Delamination Test

REFERENCE: de Charentenay, F. X., Harry, J. M., Prel, Y. J., and Benzeggagh, M. L., **"Characterizing the Effect of Delamination Defect by Mode I Delamination Test,"** *Effect of Defects in Composite Materials, ASTM STP 836,* American Society for Testing and Materials, 1984, pp. 84–103.

ABSTRACT: Delamination is an important step in the fracture of laminates. In order to characterize this phenomenon, tension opening (Mode I) tests have been carried out. Monitoring by acoustic emission allows the determination of an initiation fracture energy that is characteristic of the material bonding. Analysis by fracture mechanics of the delamination propagation shows that the resistance curves description is a very useful presentation of the delamination behavior. Results of experiments on glass-polyester, carbon-epoxy, and kevlar-epoxy are presented. The influence of moisture content on the R-curves of carbon-epoxy laminates is included.

KEY WORDS: composite materials, fracture mechanics, delamination, laminates, fatigue (materials)

The process of fracture in laminates has been considered in many papers for more than 15 years. Elementary events such as matrix cracking, interface debonding, and fiber breaking are combined in a very complicated way for the description of damage growth and fracture. Although the state of knowledge is not very high at the present time, a few principles have been firmly established. First ply failures, intralaminar transverse cracking in 90-deg layers, shear failures, characteristic damage state, delamination, etc., are now fairly well-known, and these mechanisms of fiber or ply failure describe to some extent the process of fracture in laminates.

The importance of delamination has been progressively pointed up. This phenomenon was formerly emphasized in bending as shear delamination. However, the delamination may occur at discontinuities by the coupling effect, and the

[1] Groupe "Polymères et Composites," Départment de Génie Mécanique, Université de Technologie de Compiègne, B. P. 233, 60206 COMPIEGNE, France.

work of Pipes and Pagano [1][2] has described the interlaminar stress components at the edge of laminates under tensile stress.

The description of delamination growth has been studied in a number of laminates under various stress states. Delamination is one of the few cases (perhaps the only one) where fracture mechanics can be applied aptly. The delamination crack is well-defined and relationships between crack length, stress level, and specimen geometry can be laid down. These relationships can be applied to monotonic or fatigue delamination growth. Fracture mechanics can thus be applied to the problem of delamination defects that may occur either during molding or by mechanical damage, such as impact. This paper emphasizes the use of fracture mechanics testing for defining delamination parameters. After reviewing the delamination tests already published in other papers, one shall describe how a thorough interpretation of Mode I testing can give a better knowledge of delamination growth.

Previous Studies

Delamination may occur either by shear (in bending) or by cleavage. Shear delamination is a Mode II crack propagation and cleavage delamination is a Mode I crack propagation in terms of fracture mechanics.

The edge delamination has been shown to occur by the interlaminar stress that appears at the discontinuities (edge, hole, etc.). Several papers have been devoted to the description of stress evolution in the vicinity of the edges of a tensile coupon [1,2,3]. The importance of stacking sequences has been emphasized. For instance, the normal stress, σ_z, can be either in compression or in tension depending on the ply sequence [4,5]. These studies show that edge delamination is a mixed-mode crack propagation (cleavage and shear, Mode I and Mode II) and that the Mode I test is an important part of the definition of delamination behavior. Since this cleavage is very easy to carry out, several authors have published studies on this delamination test.

Mode I Delamination Tests

The simple double cantilever beam (DCB) specimen used by several authors [6–12] is presented schematically in Fig. 1. The defect is introduced during molding by a thin nonadhesive film. The compliance method is used for the determination of fracture energy, G_c

$$G_c = \frac{P_c^2}{2\,w}\frac{dC}{da}$$

[2] The italic numbers in brackets refer to the list of references appended to this paper.

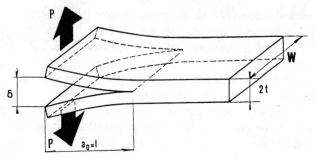

FIG. 1—*Mode I delamination test.*

where

P_c = critical load of delamination,
C = compliance,
w = width of specimen, and
a = length of defect.

The dC/da is evaluated by a set of specimens with various lengths of defects. Devitt et al [13] has introduced corrections for thin specimens that severely deform before delamination takes place. Only a few authors have noticed differences between initiation and propagation fracture energy [8,9,10,12]. In order to facilitate interlaminar fracture testing, Bascom et al [14] has utilized a width tapered DCB specimen like Brussat et al [15] for adhesion testing. In this constant dC/da specimen, the load during delamination is independent of a.

Mode II Delamination Tests

Pure Mode II tests (sliding or shear) are very difficult to work out. de Charentenay et al [9] and Russell and Street [12] have proposed tests of bending on partially delaminated beam. Conditions of loading and specimen geometry have to be carefully optimized. Development of such tests in a number of laboratories should appear soon.

Mixed-Mode Delamination Tests

The study of shear lap tests carried out by Wilkins et al [11] has been shown to be a mixed Mode I and II test (25% Mode I for the specimen used). This is calculated by finite element method. Also, the edge delamination of (±30, ±30, 90, 90)$_s$ by tension test is shown to be a mixed-mode delamination test [16,17]. Since in this experiment the strain energy release rates, G_I and G_{II}, are independent of the delamination size, the determination of total G_c needs only the measurement of the strain at delamination start.

Mechanics of Mode I Tests

The application of linear elastic fracture mechanics on Mode I delamination is based on compliance method (Fig.1)

$$G_c = \frac{P_c^2}{2\,w}\frac{dC}{da}$$

(1)

where

$$C = \frac{\delta}{P}, \text{ and}$$

δ = displacement.

The compliance function of crack length, a, from the simple beam theory, is

$$C = \frac{2a^3}{3EI}$$

(2)

where

E = modulus, and
I = moment of inertia.

However, the exact description is slightly different than the simple beam theory due to the finite length of the beam and some rotation components due to the thickness. This can be taken into account by either a polynomial coefficient [18,19] or more simply by replacing the Exponent 3 by an n-value [20,21]

$$C = \frac{a^n}{h}$$

(3)

The coefficients, n and h, are determined by experiment on a set of specimens with various a-values. The compliance, C, measured at low displacement is utilized in a log-log plot

$$\log C = n \log a - \log h$$

(4)

The fracture energy, G_c, can thus be evaluated

$$G_c = \frac{n\,P_c\,\delta_c}{2a\,w}$$

(5)

where subscript c means critical.

The calculation by the same procedure of strain energy release rate G (available elastic energy) in the function of a for a given δ shows that, if G_c is considered to be constant, delamination must be stable throughout the experiment (Fig. 2).

This method allows us to determine easily critical fracture energy for delamination. However, as shown by a few papers [9,10,12], the fracture energy does not remain constant when delamination grows. The increase of G_c during growth is well described by the resistance curves (R-curves) that are the plot of instantaneous G_c in the function of a. The determination of R-curves may be easily done by the compliance method:

(1) The a can be measured by compliance method using Eq 3 either by a few unloadings during the experiment (Fig. 3a) or by drawing a straight line from the origin in the case of small residual displacement (Fig. 3b).

(2) Knowing the coefficients n and h, G can be calculated from Eqs 3 and 5

$$G_p = \frac{n}{2\ w\ h^{1/n}} \{P_p^{n+1}\ \delta_p^{n-1}\}^{1/n}$$

where subscript p means propagation.

Thus the construction of R-curves only needs the calibrated function $C = f(a)$

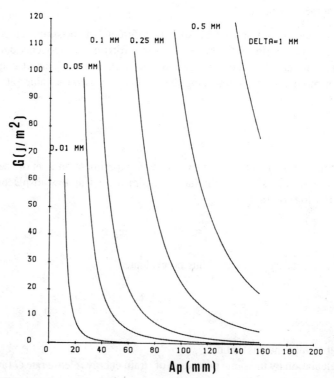

FIG. 2—*Strain energy release rate versus crack length at constant displacement* δ.

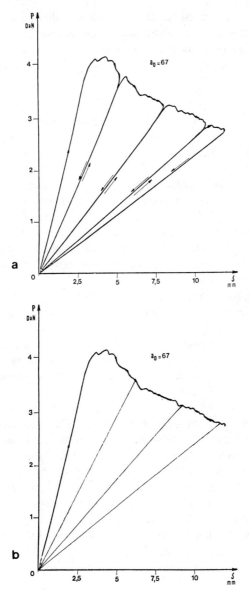

FIG. 3—*Construction of R-curves. (a) Determination of compliance for a set of points by unloading and reloading. (b) Determination of compliance for a set of points by drawing straight lines from the origin.*

and the load-displacement curve. However the calculated crack length, a_p, is slightly different than the true delaminated crack because the compliance is also modified by the damage at the crack tip. Thus, the length, a_p, is larger than the true crack length, a, but smaller than the length, a, plus the damage length [22].

Experiments

Several materials have been tested in Mode I delamination. The glass fiber reinforced plastics (GFRP) are glass woven fabrics with polyester resin (Synrès S.A). The carbon reinforced composites are made of T300 Toray fibers and epoxy resin either from Narmco (N5208) or from Ciba Geigy (BSL914). Some of the prepregs are from Brochier. The carbon-epoxy and kevlar-epoxy, both unidirectional, were molded by Société Nationale Industrielle Aérospatiale and Etablissement Technique Central de l'Armement. During the molding, a thin film (Teflon for polyester, and a piece of molding bag or vac-pak for the carbon and kevlar-epoxy) is introduced in the middle of the plates. Samples containing various lengths of interlaminar defect are cut from the molded plates. Dimensions of the tested samples are shown in Table 1.

Some of the carbon-epoxy samples were moisture conditioned by immersion in water at 90°C during a period calculated by a program using Fick's law and a finite difference method.

The moisture content through the thickness was homogenized in stainless steel containers at 90°C during a period calculated by the same program. After opening the container, the samples were stored in a desiccator over various saturated salt solutions that control the moisture content in the atmosphere [23].

The loading of the samples is executed by means of two hinges. One part of both hinges is glued on both sides of the sample. The other part is clamped in the testing machine jaws. The delamination test is carried out at a displacement rate of 2 mm/min. An X-Y recorder plots the load-displacement curves.

Acoustic emission was utilized for monitoring the microscopic process of fracture. The signal of a piezoelectric transducer (300 kHz resonant frequency) is amplified (90 dB), filtered (frequency lower than 50 kHz are filtered out), and fed to a discriminator that transforms burst signals into counts (LEANORD-CGR). The threshold of the discriminator was fixed to 0.5 V. However, we have checked, and that variation of the threshold value from 0.4 to 0.7 V does not influence the counts rate recording.

Results and Discussion

Initiation

Typical load and acoustic emission counts rate versus displacement curves are shown for carbon-epoxy (Fig. 4), glass-polyester (Fig. 5), and kevlar-epoxy (Fig. 6). From the first part of these curves, the compliance can be measured and the coefficients, n and h, determined (Fig. 7, for example).

TABLE 1—*Tested composites.*

Materials	Resin	Fiber	a_o, mm	Width, mm	Thickness, mm	Length, mm
Glass-polyester (cloth, 270 g/m²)	Stratyl 116 Rhone-Poulenc	roving	25–90	31	3.2	100 to 150
Glass-polyester (cloth, 830 g/m²)	Norsodyne 294 T CDF chimie	roving	25–88.5	31	3.2	
Glass-polyester unidirectional, 420 g/m²	Stratyl 116 Rhone-Poulenc	unidirectional	22–96.5	31	3.3	
Carbon-epoxy unidirectional, T300-5208	Epoxy: N5208 Narmco Div. Whittaker Corp.	T300 (Toray Industry)	39–103 40–100	20	3.2 5.3	
Carbon-epoxy unidirectional, T300 Brochier	Epoxy: Brochier NCHR 1424	T300	23–76	30	2	
Carbon-epoxy unidirectional, T300-BSL914	Epoxy: BSL914 Ciba-Geigi	T300	40–100	20	5.3	
Kevlar-epoxy unidirectional, Brochier	Epoxy: Brochier BK 1424	PRD 49	22–79	30	2	
Glass-epoxy unidirectional, Brochier	Epoxy: Brochier BR 1424	R	24–74	30	2	

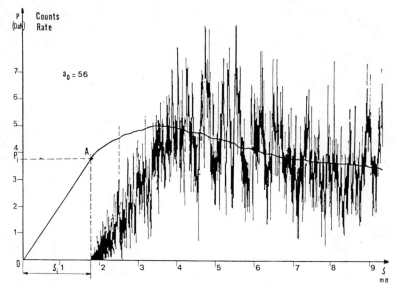

FIG. 4—*Load and acoustic emission counts rate versus displacement for T300-N5208 carbon-epoxy.*

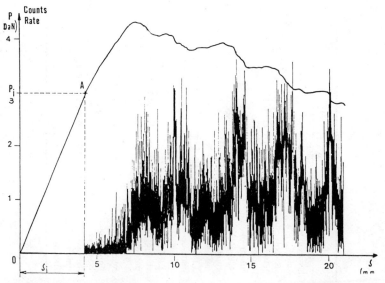

FIG. 5—*Load and acoustic emission counts rate versus displacement for glass-polyester.*

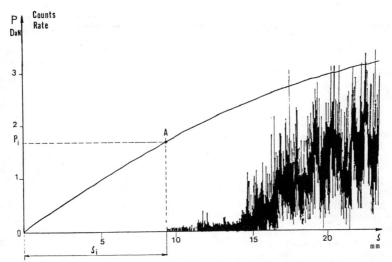

FIG. 6—*Load and acoustic emission counts rate versus displacement for kevlar-epoxy.*

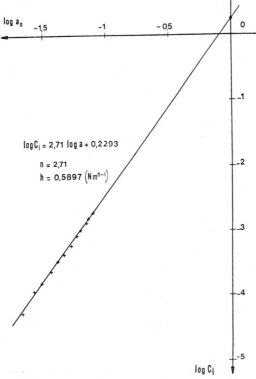

$$\log C_i = 2{,}71 \log a + 0{,}2293$$

$$n = 2{,}71$$

$$h = 0{,}5897 \ (Nm^{n-1})$$

FIG. 7—*Log-log plot of compliance versus initial defect length. Determination of the constants,* n *and* h.

TABLE 2—*Initiation fracture energy in delamination.*

Materials	a_o, mm	G_i, J/m^2	Standard Deviation
Glass-polyester (cloth, 270 g/m^2)	25–90	165	14.65
Glass-polyester (cloth, 830 g/m^2)	25–88.5	102.9	6.03
Glass-polyester unidirectional, 420 g/m^2	22–96.5	133.46	5.3
Carbon-epoxy unidirectional, T300-5208	39–103	59.8	8.46
Carbon-epoxy unidirectional, T300 Brochier	23–76	45.37	3.69
Kevlar-epoxy unidirectional, Brochier	22–79	118.85	18.8
Glass-epoxy unidirectional, Brochier	24–74	165.13	2.44

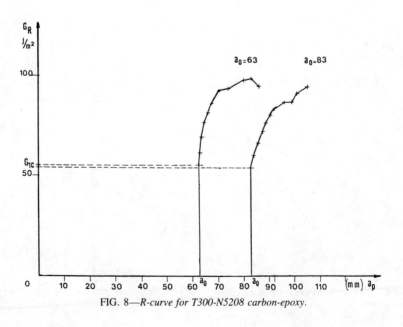

FIG. 8—*R-curve for T300-N5208 carbon-epoxy.*

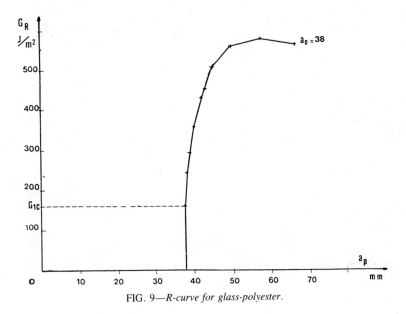

FIG. 9—*R-curve for glass-polyester.*

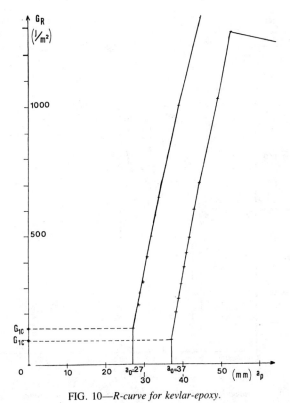

FIG. 10—*R-curve for kevlar-epoxy.*

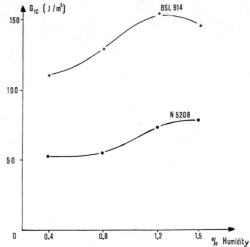

FIG. 11—*Initiation fracture energy versus moisture content for T300-N5208 and T300-BLS914 carbon-epoxy.*

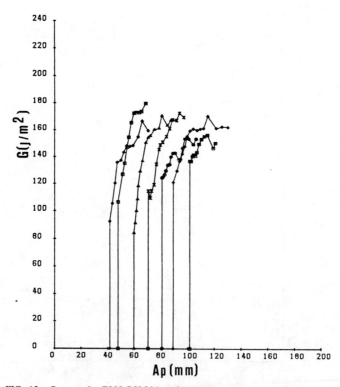

FIG. 12—*R-curve for T300-BSL914 carbon-epoxy, moisture content = 0.4%.*

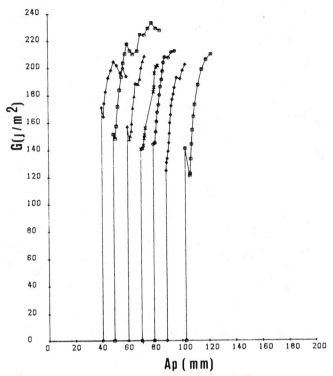

FIG. 13—*R-curve for T300-BSL914 carbon-epoxy, moisture content* = *1.5%.*

In a few experiments, the first part of the load-displacement, which should be a straight line, shows a slight positive curvature (see Fig. 3, for example). This is due to the non-linearity introduced by high displacement in bending [13]. In these few cases, although this is not strictly correct, we have used as a compliance value the one determined by the straight line between the origin and the point of acoustic emission initiation.

Acoustic emission onset defines very accurately the initiation of microscopic damaging at the crack tip before macroscopic delamination. Results on several specimens are presented in Table 2. The initiation fracture energy, G_i, is independent of the initial delamination length. G_i is thus a material parameter and defines an intrinsic microcracking fracture energy.

The measured values appear to be fairly low compared with results of experiments that did not check damage or delamination initiation point [13,14]. The neat resin values were not measured in our case but other results show higher fracture energy for common epoxy or polyester resin. This can be explained by considering that the first damage occurs at the fiber-matrix interface that is less resistant than the matrix itself. Attention must be paid to the particularly low value of the T300-N5208 carbon-epoxy interface.

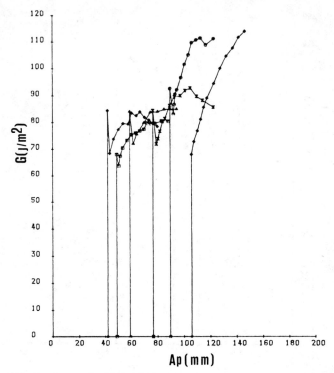

FIG. 14—*R-curve for T300-N5208 carbon-epoxy, moisture content = 0.4%.*

Propagation

Typical R-curves for the Mode I test on the three types of laminates are shown on Figs. 8 (carbon-epoxy), 9 (glass-polyester), and 10 (kevlar-epoxy). These R-curves show that, after initiation, the fracture energy grows very rapidly and becomes somewhat constant at a level that is much higher than the initiation fracture energy. Results published in various papers are located between the initiation fracture energy and the maximum propagation fracture energy. The determination of the R-curves that includes the different steps of delamination is much more informative for the characterization of materials.

The comparison of the three curves (Figs. 8, 9, and 10) on the three laminates is an illustration of the different delamination behaviors. In carbon-epoxy, the increase of fracture energy is lower than in glass-polyester indicating that the damage mechanism by microscopic failure around the delamination crack tip is very different from one material to the other. The huge increase of fracture energy in kevlar-epoxy can be due to the delamination of the fibers that occur in the adjacent planes of reinforcement.

It is interesting to note that these damage developments as described by R-curves are in the same order as the impact resistance measured by a falling

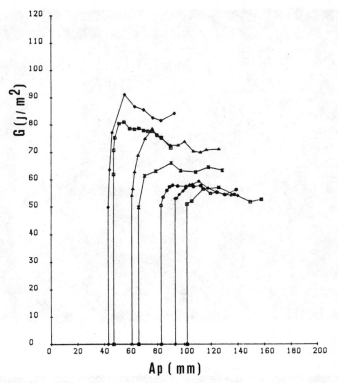

FIG. 15—*R-curve for T300-N5208 carbon-epoxy, moisture content = 1.5%.*

weight experiment. Damage is a known mechanism that enhances impact resistance.

In order to check further the significance and validity of the R-curves, we
have performed a set of experiments on T300-N5208 and T300-BSL914 carbon-
epoxy unidirectional laminates in various moisture contents. Delamination tests
were conducted for each material and moisture content using a set of six to seven
specimens with various lengths of interlaminar defects. The initiation fracture
energy was slightly sensitive to the moisture content in both cases (Fig. 11).
The result obtained here for T300-N5208 is consistent with the value of 60
J/m² measured on the same material but not from the same molding batch.
Figures 12 and 13 show the R-curves for the two extreme moisture contents of
the T300-BSL914. In this case, the R-curves are not very much different for the
two states, and the influence of initial defect length is low. However, the R-
curves determined on the T300-N5208 are very sensitive not only to the moisture
content but also to the length of initial defect (Figs. 14 and 15). At low moisture
content, the R-curves are lower when the initial defect length is larger and some
of the R-curves present negative slope. At high moisture content, the R-curves
are higher when the initial defect length is larger.

FIG. 16—*SEM photomicrograph of delaminated surface of T300-BSL914 carbon-epoxy, moisture content = 0.4%.*

FIG. 17—*SEM photomicrograph of delaminated surface of T300-BSL914 carbon-epoxy, moisture content = 1.5%.*

FIG. 18—*SEM photomicrograph of delaminated surface of T300-N5208 carbon-epoxy, moisture content = 0.4%.*

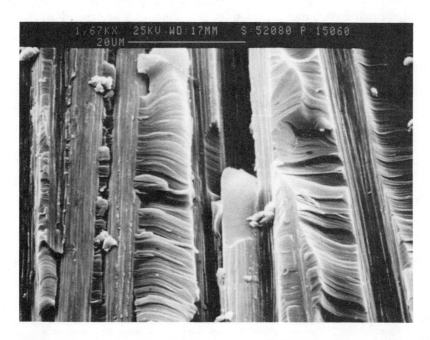

FIG. 19—*SEM photomicrograph of delaminated surface of T300-N5208 carbon-epoxy, moisture content = 1.5%.*

This kind of variation of delamination growth parameters with the initial defect length has also been verified on glass-polyester [10]. Thus, in this case, fracture mechanics does not hold. The influence of local shear stress components for high displacement, δ, arising at a large length of delamination, may modify the damaging mechanism and hence the propagation fracture energy. Since moisture plasticizes the epoxy matrix, this influence may be moisture dependent.

Scanning electron microscope (SEM) photomicrography does not show significant differences between the two moisture contents for both materials (Figs. 16 to 19). However, differences between the two materials are clearly shown: the T300-BSL914 undergoes fracture mainly in the matrix, while the T300-N5208 shows interfacial debonding. This is consistent with the lower initiation fracture energy of this material.

Conclusion

Mode I delamination tests must be carefully monitored and interpreted in order to get precise descriptions of delamination processes and accurate parameters for an objective characterization. The initiation fracture energy has defined the onset of damage by microscopic failure of matrix or interface and is a characteristic material parameter. The resistance curves described development of the damage during delamination in which a large part of fracture energy arises from multiple cracking and debonding. Improving the interlaminar resistance should take into account the two stages of the delamination process. A good material must not only have a high initiation (or first failure) fracture energy but also develop extensive damage before and during the delamination propagation.

References

[1] Pipes, R. B. and Pagano, N. J., *Journal of Composite Materials*, Vol. 4, 1970, p. 538.
[2] Wang, S. S. and Choi, I., *Journal of Applied Mechanics*, Part 1 and Part 2, Vol. 49, 1982, p. 541.
[3] Herakovich, C. T., Nagarkar, A., and O'Brien, D. A., "Failure Analysis of Composite Laminates with Free Edges," *Modem Developments in Composite Materials and Structures*, J. R. Vinson, Ed., American Society of Mechanical Engineers, 1979.
[4] Pagano, N. J. and Pipes, R. B., *Journal of Composite Materials*, Vol. 5, 1971, p. 50.
[5] Herakovich, C. T., "On the Relationship Between Engineering Properties and Delamination of Composite Laminates," report of the Virginia Polytechnic Institute, No. VPI-E-81-2, 1981.
[6] McKenna, G. B., Mandell, J. F., and McGarry, F. J., Society of the Plastic Industry, Annual Technical Conference, 1974, Section 13 C.
[7] Scott, J. M. and Phillips, D. C., *Journal of Materials Science*, Vol. 10, 1975, p. 551.
[8] Bethmont, M., "Etude par Emission Acoustique de Matériaux Composites Stratifiés," Thèse, Université de Compiègne, 1977.
[9] de Charentenay, F. X., Bethmont, M., Benzeggagh, M., and Chretien, J. F. in *Mechanical Behaviour of Materials* (ICM 3), K. J. Miller, and R. F. Smith, Eds., Vol. 3, 1979, p. 241.
[10] de Charentenay, F. X. and Benzeggagh, M., "Fracture Mechanics of Mode I Delamination in Composite Materials," 4th International Conference on Composite Materials, A. R. Bunsell, Ed., Vol. 1, 1980, p. 186.
[11] Wilkins, D. J., Eisenmann, J. R., Camin, R. A., Margolis, W. S., and Benson, R. A., "Characterizing Delamination Growth in Graphite-Epoxy," *Damage in Composite Materials*, ASTM STP 775, American Society for Testing and Materials, 1982.

[12] Russell, A. J. and Street, K. N. in *Progress in Science and Engineering of Composites* (ICCM IV), T. Hayashi, K. Kawata, and S. Umekawa, Eds., 1982, p. 279.
[13] Devitt, D. F., Scharery, R. A., and Bradley, W. L., *Journal of Composite Materials,* Vol. 14, 1980, p. 270.
[14] Bascom, W. D., Bitner, J. L., Moulton, R. J., and Siebert, A. R., *Composites,* Jan. 1980, p. 9.
[15] Brussat, T. R., Chiu, S. T., and Mostovoy, S., "Fracture Mechanics for Structural Adhesive Bonds," IR 27614-9 to 11, Air Force Materials Lab., Wright Patterson Air Force Base, Ohio, 1977, 1978.
[16] Rybicki, E. F., Schmueser, D. W., and Fox, J., *Journal of Composite Materials,* Vol. 11, 1977, p. 470.
[17] O'Brien, T. K., Johnston, N. J., Morris, D. H., and Simonds, R. A. in *Journal,* Society for the Advancement of Materials and Process Engineering, July–Aug. 1982, p. 8.
[18] Kanninen, M. F., *International Journal on Fracture,* Vol. 9, 1973, p. 83.
[19] Gillis, P. P. and Gilman, J. J., *International Journal of Solids Structures,* Vol. 12, 1976, p. 13.
[20] Gilman, J. J., *Journal of Applied Physics,* Vol. 31, 1960, p. 2208.
[21] Berry, J. P., *Journal of Applied Physics,* Vol. 34, 1963, p. 62.
[22] Benzeggagh, M., "Application de la Mécanique de la Rupture aux Matériaux Composites: Exemple de la Rupture par Délaminage d'un Stratifié," Thèse, Université de Compiègne, 1980.
[23] Shirell, C. D. in *Advanced Composite Materials—Environmental Effects, ASTM STP 658,* J. Vinson, Ed., American Society for Testing and Materials, Philadelphia, 1978, p. 21.

J. Whitney[1] and Charles E. Browning[1]

Materials Characterization for Matrix-Dominated Failure Modes

REFERENCE: Whitney, J. M. and Browning, C. E., **"Materials Characterization for Matrix-Dominated Failure Modes,"** *Effects of Defects in Composite Materials, ASTM STP 836*, American Society for Testing and Materials, 1984, pp. 104–124.

ABSTRACT: Viable test methods that may be used in characterizing matrix-dominated composite properties are assessed. Special emphasis is placed on matrix cracking and delamination. Both strength characterization and fracture mechanics test methods are considered. Transverse tension, in-plane shear, interlaminar shear, and a laminate test for *in situ* first-ply failure are the strength tests considered. Fracture mechanics test methods discussed include the double-cantilever-beam test, a 90-deg center-notch tension test, and a free-edge-delamination tension test.

KEY WORDS: composite materials, fatigue (materials), fracture mechanics, delamination, matrix-related properties, first-ply failure, tension tests

Standard test methods currently in use for characterizing new or improved matrix resins for application in high-performance fiber-reinforced composites give limited information concerning matrix-dominated failure modes. Since matrix cracking and delamination are of key concern in assessing "toughness" in composite laminates, current test philosophy must be modified to accommodate methods that interrogate the materials resistance to transverse ply cracking and delamination.

Among the standard test methods utilized, only 90-deg unidirectional tension, unidirectional in-plane shear (±45-deg tension, 10-deg off-axis tension, and rail shear), and short beam shear give any information concerning the matrix-dominated failure modes under consideration. Transverse tension and in-plane shear are often used to set design allowables on matrix-dominated first-ply failure, while the short beam shear test is the only interlaminar test performed.

The matrix also provides support to the fiber so that 0-deg compression can be sensitive to matrix properties, especially at elevated temperatures. Since compression failure is usually stability related (fiber micro-buckling, local crippling due to fiber waviness, and gross buckling), test data are a function of

---

[1] Materials research engineer, Nonmetallic Materials Division, Materials Laboratory, Air Force Wright Aeronautical Laboratories, Wright-Patterson Air Force Base, Ohio 45433.

specimen geometry and support conditions. This makes it very difficult to relate resin properties directly to unidirectional compression strength. Thus, compression tests are not considered in the present study.

Two generic approaches are available for interrogating a composite material's resistance to matrix cracking and delamination. The first approach involves a detailed stress analysis used in conjunction with a failure criterion to predict, and measure experimentally, the onset of matrix cracking and delamination. In the second approach, classical techniques of linear elastic fracture mechanics are applied to characterizing matrix cracking and delamination. Current trends point toward the first approach for characterizing matrix cracking and the second approach for delamination studies.

In this paper, viable test methods that may be used in determining composite toughness, with special emphasis on matrix cracking and delamination, are assessed. Both strength characterization and fracture mechanics test methods are considered.

In order to assess the sensitivity of the various test methods, the corresponding composite properties, to matrix stress-strain behavior, two model resin systems are utilized extensively in the experimental data presented. Typical tensile stress-strain curves for these two resin systems are illustrated in Fig. 1. The AF-R-E350 resin is a brittle system consisting of MY-720 (a tetra-functional epoxy from Ciba-Geigy) and DDS (diaminodiphenylsulfone curing agent from Ciba-Geigy) custom mixed to Air Force directions. Union Carbide's polysulfone represents a ductile matrix.

Tension dogbone specimens approximately 4.1 mm (0.16 in.) wide and 51 mm (2 in.) long in the gage section were utilized in obtaining the AF-R-E350 tensile data. An extensometer was used for measuring strain. The polysulfone stress-strain curve was obtained from data supplied by Union Carbide in which a dogbone specimen was utilized in conjunction with ASTM Test for Tensile Properties of Plastics (D 638-77a). The initial tensile modulus of AF-R-E350 is

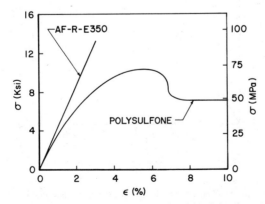

FIG. 1—*Typical tensile stress-strain curves for brittle and ductile resin systems.*

approximately 3.46 GPa (500 ksi) and the polysulfone approximately 2.76 GPa (400 ksi).

Shear stress-strain data for these two resin systems are shown in Fig. 2. These stress-strain curves were obtained from a three-rail fixture [1][2] using a 152-mm (6-in.) square plate.

Strength Characterization

In this section, the conventional tests for measuring transverse tensile strength, in-plane shear strength, and interlaminar shear strength are discussed. These are the tests normally utilized for measuring matrix-dominated strength properties. In addition, consideration is given to tensile loading of a laminate for the purpose of determining *in situ* transverse-ply failure.

Transverse Tension

Strain to failure as determined from a 90-deg unidirectional tension test is often the basis for determining engineering strength in a laminated composite material. In particular, if first-ply failure is assumed to determine laminate strength, then for composites containing 90-deg plies, transverse tension strength becomes a measure of laminate strength.

Stress-strain curves are illustrated in Fig. 3 for AS-1/AF-R-E350 and AS-1/ polysulfone unidirectional composites. These tests were performed on 25.4-mm

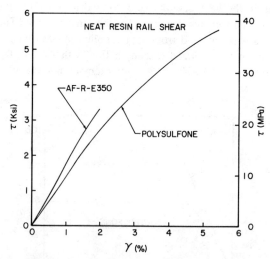

FIG. 2—*Shear stress-strain curves for brittle and ductile resin systems.*

[2] The italic numbers in brackets refer to the list of references appended to this paper.

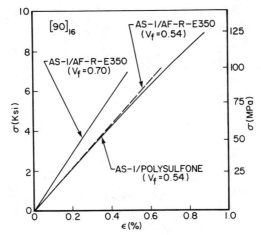

FIG. 3—*Transverse tensile stress-strain curves.*

(1-in.) wide, straight-sided tensile coupons. The tests were performed in accordance with ASTM Test for Tensile Properties of Oriented Fiber Composites (D 3039-76).

Although the more ductile polysulfone composite has an increased strain to failure compared to the brittle AF-R-E350 matrix composite, the increase is not in proportion to neat resin strain capability. Yielding of the polysulfone in neat resin form leads to necking of tension specimens at high strain levels. Since the resin is under biaxial stress in the composite, necking is prevented. Perhaps biaxial stress-strain data is more indicative of resin performance in the composite.

When making a comparison between composites containing different matrix systems, the effect of fiber volume content must be considered. For the composites in Fig. 3, a large difference in volume content between the two composites under consideration is observed. In order to make a comparison on an equal fiber volume basis, the AS-1/AF-R-E350 data was extrapolated to a fiber volume fraction of $V_f = 0.54$. The extrapolated results appear as the dotted line in Fig. 3. This extrapolation was obtained by assuming the strain to failure for the brittle matrix composite is determined by the maximum strain concentration factor in the matrix, which is a function of fiber volume fraction. The strain concentration factor can be estimated by breaking the repeating element of the fiber packing geometry into thin strips, as illustrated at the top of Fig. 4 for hexagonal packing. This approach has been discussed in detail by Chamis [2] and leads to the strain concentration expression

$$\frac{\epsilon_m}{\epsilon_T} = \left[1 - \left(1 - \frac{E_m}{E_{fT}} \right)\left(\frac{k\,V_f}{\pi} \right)^{1/2} \right]^{-1} \tag{1}$$

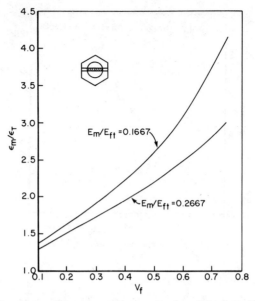

FIG. 4—*Transverse strain concentration as a function of fiber volume fraction.*

where

ϵ_m = average strain in matrix along the center strip (see Fig. 4),

ϵ_T = average transverse strain in composite,

E_m = modulus of the matrix,

E_{fT} = transverse modulus of the fiber, and

k = constant that depends on fiber packing geometry.

For square packing, $k = 4$, and for hexagonal packing, $k = 3.464$. The strain concentration factor is plotted as a function of fiber volume fraction in Fig. 4 for two ratios of E_m/E_{fT}. The curve for $E_m/E_{fT} = 0.1667$ corresponds to $E_m = 3.46$ GPa (500 ksi), and $E_m/E_{fT} = 0.2667$ corresponds to $E_m = 5.53$ GPa (800 ksi). In both cases, $E_{fT} = 20.7$ GPa (3×10^6 psi).

Since the neat resin tensile stress-strain curve is linear to failure, it is assumed that the transverse composite stress-strain curve is linear for all values of V_f. Ultimate strength can then be determined at $V_f = 0.54$ from the transverse tensile modulus of the composite, E_T, and the strain at failure as extrapolated from Eq 1. The Halpin-Tsai equations [3] are used to extrapolate E_T from $V_f = 0.70$ to $V_f = 0.54$. In particular

$$E_T = \frac{E_m[E_{fT}(1 + V_f) + E_m(1 - V_f)]}{E_{fT}(1 - V_f) + E_m(1 + V_f)} \qquad (2)$$

Using this approach, the dotted line in Fig. 2 yields little change in the ultimate strength while the ultimate strain increases from 0.485% at $V_f = 0.70$ to 0.646% at $V_f = 0.54$. Thus, volume percent of fiber must be considered when comparing the effect of matrix properties on composite properties. For the AS-1/AF-R-E350, $E_T(V_f = 0.70) = 10$ GPa (1.45×10^6 psi) and $E_T(V_f = 0.54) = 7.53$ GPa (1.09×10^6 psi). For the AS-1/polysulfone, $E_T = 7.32$ GPa (1.06×10^6 psi.)

In-Plane Shear

Shear stress-strain curves are shown in Fig. 5 for unidirectional composites. Rail shear data are obtained from a three-rail fixture [1] using a 152 mm (6 in.) square plate. For shear data derived from a ±45-deg tensile coupon, specimens were straight-sided tensile coupons 25.4 mm (1 in.) wide tested in accordance with ASTM Test D 3039-76. Data is reduced in accordance with ASTM Recommended Practice for In-plane Shear Stress-Strain Response of Unidirectional Reinforced Plastics (D 3518-76). Although correlation can be observed between initial shear stress-strain response of the rail shear and ±45-deg tensile coupon, the rail shear fails prematurely with the failure propagating from the bolt holes. The ±45-deg tensile coupon results in an unrealistically high shear strength due to the fact that some of the fibers run through the grip section and are clamped. The rail shear test may, however, provide a sufficient portion of the stress-strain curve to be useful from an engineering standpoint. In Fig. 5, the longitudinal shear stress is denoted by τ_{LT} and the longitudinal shear strain by γ_{LT}. The initial shear modulus for AS-1/AF-R-E350 is $G_{LT} = 5.53$ GPa (800 ksi) and for the AS-1/polysulfone, $G_{LT} = 3.87$ GPa (560 ksi).

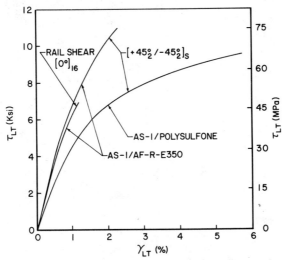

FIG. 5—*Shear stress-strain curves for unidirectional composites.*

The nature of the shear stress-strain curves makes it difficult to assess the effect of matrix properties on the in-plane shear response of a unidirectional composite.

Interlaminar Shear

Interlaminar shear strength of a unidirectional composite is usually determined by a short beam shear test [ASTM Test for Apparent Horizontal Shear Strength of Reinforced Plastics by Short Beam Method (D 2344-76)]. For high-performance composite materials, the method is often used in conjunction with a thin beam (12 to 16 plies) under three-point bending. In order to induce a shear failure, the span-to-depth ratio is nominally 4:1. Such thin beams rarely, if ever, produce an interlaminar shear failure. A typical failure mode for such specimens is illustrated in Fig. 6, where a photomicrograph of the top section of a failed 0-deg, 16-ply, AS-1/AF-R-E350 beam reveals localized buckling adjacent to the load point (loading was applied to the right of the buckled region). For a 50-ply beam of the same material, a horizontal split is observed. However, the photomicrograph in Fig. 7 of the failed beam reveals a vertical crack that initiated prior to the horizontal split. The load was applied to the left of the vertical crack. Some localized buckling can also be observed behind the vertical crack. These failure modes occur in regions where the stress distribution is very complex.

FIG. 6—*Photomicrograph of 16-ply, AS-1/AF-R-E350 short-beam shear specimen.*

FIG. 7—*Photomicrograph of 50-ply, AS-1/AF-R-E350 short-beam shear specimen.*

The shear stress distribution at three different cross sections of the 16-ply beam is illustrated in Fig. 8. Note that classical beam theory is accurate over a very limited segment of the beam. This stress distribution was obtained from a plane stress Fourier series solution in which the concentrated loads were modeled as uniform stresses distributed over a very small length of the beam. The axis

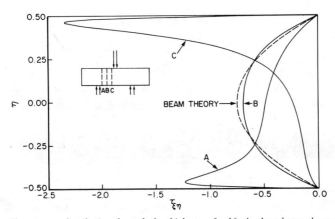

FIG. 8—*Shear stress distribution through-the-thickness of a 16-ply short-beam shear specimen.*

system is at the centerline of the beam. The thickness coordinate, η, is normalized by the beam thickness, h, while the shear stress is normalized by the maximum bending stress.

Horizontal splitting can also be produced on graphite/epoxy beams under four-point bending, if the span-to-depth ratio is reduced to 16:1 from the usual 32:1 [4]. A photomicrograph of a failed 0-deg, 24-ply, AS-1/AF-R-E350 beam under four-point bending is shown in Fig. 9. Again the horizontal split is preceded by a vertical crack. In this case, the horizontal split initiates near the tip of the vertical crack and begins moving toward the center of the beam and then reverses direction, propagating catastrophically to the end of the beam. In Fig. 9, the load was applied to the left of the vertical crack.

The complex failure modes along with the complex stress distribution make interlaminar failures in beam specimens difficult to interpret. Furthermore, horizontal failures are difficult to induce in thermoplastic matrix composites such as the AS-1/polysulfone material. These restrictions reduce considerably the usefulness of these interlaminar beam tests.

In Situ *Transverse Strength*

A 90-deg unidirectional tension test gives useful information concerning the transverse strain capability of the material under consideration. However, a

FIG. 9—*Photomicrograph of 24-ply, AS-1/AF-R-E350 four-point beam shear specimen.*

multidirectional laminate containing 90-deg plies may provide a more realistic approach to both measuring and defining *in situ* first-ply failure.

By using a multidirectional laminate, the results may also be more realistic for engineering purposes. In particular, a 90-deg tension test provides initial transverse-ply failure data only, while a multidirectional laminate allows one to observe multiple cracks in transverse plies and assess their effect on laminate mechanical behavior. Such constraint plies have several important effects. In addition to offering resistance to crack growth in the transverse plies perpendicular to the load direction, they also reduce the effect of surface flaws on first-ply failure. It has been experimentally observed by Flaggs and Kural [5] that *in situ* transverse strain to failure, including the effects of residual thermal strains, can be almost twice the failure strain of a transverse, unidirectional tensile coupon.

This phenomenon of *in situ* transverse reinforcement is illustrated in Fig. 10, where the tensile stress-strain curve for a Hercules' AS-1/3502 laminate with the stacking geometry [45/90/ − 45/90/45/90/ − 45/90]$_s$ is shown. Transverse crack density in the center 90-deg ply as a function of axial strain level is also illustrated. Note that the onset of any measurable transverse crack density closely correlates with the knee in the stress-strain curve that occurs at approximately 0.8% axial strain. Although this strain level is likely to be a function of the ply orientations adjacent to the 90-deg plies and the number of 90-deg plies stacked together [5], the results in Fig. 10 are informative. Thus, any materials characterization scheme should consider including a laminate tension test that will interrogate *in situ* transverse-ply strength. The data in Fig. 10 was obtained from a 25.4-mm (1-in.) wide, straight-sided tensile coupon tested in accordance with ASTM Test D 3039-76. Crack density was determined by stopping the test at

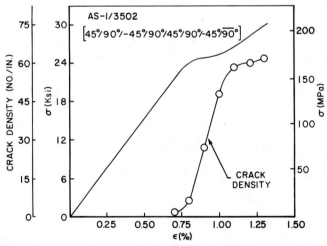

FIG. 10—*Tensile stress-strain curve and transverse crack density for an angle-ply laminate.*

various load levels, removing the specimen, placing it under a microscope, and counting the number of cracks along the edge of the specimen. The edges of the specimen were polished prior to testing.

Interlaminar Fracture Mechanics Characterization

Since composite laminates do not commonly develop through-the-thickness cracks as a prelude to catastrophic failure, fracture mechanics does not apply to composite materials in the classical manner of metallic materials. However, in the case of delamination, fracture mechanics appears to be a natural approach. In particular, the failure process can be characterized by the propagation of a single crack.

Fracture due to both interlaminar tension and interlaminar shear are of concern. Thus, characterization of delamination should include both interlaminar Mode I and Mode II along with mixed-mode type tests. To date, interlaminar Mode I has received the most emphasis and is the focus of attention in this paper. Consideration is given to the double-cantilever-beam test, 90-deg center-notch coupon, and the free-edge delamination tensile coupon.

The Double-Cantilever-Beam Test

A double-cantilever beam specimen is illustrated in Fig. 11 along with the data reduction scheme. The specimen is 229 mm (9 in.) long and 25.4 mm (1 in.) wide. Standard thickness is 24 plies. A 25.4-mm (1-in.) long teflon strip is placed in the center of the laminate during fabrication in order to create a starter crack. Load introduction is achieved through the use of extruded aluminum "T" end tabs bonded to the specimen at room temperature. These tabs contain holes for connecting pins that attach the specimen to the loading fixtures, and slots

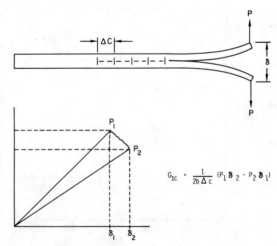

$$G_{Ic} = \frac{1}{2b \Delta c} (P_1 \delta_2 - P_2 \delta_1)$$

FIG. 11—*Double-cantilever-beam test.*

for adjustable screws that attach to an extensometer for measuring deflection [6].

Edges of the specimen are marked at 12.7-mm (0.5-in.) increments with a silver-leaded pencil for visual crack growth determinations. The initial increment of loading produces a pop-in phenomenon, resulting in erratic loading and unloading response. In addition, this first increment produces a natural crack from which a critical strain energy release rate, G_{Ic}, can be determined on further increments of crack growth. Subsequent tests are then performed on the same specimen for each 12.7-mm (0.5-in.) increment marked on the specimen. The specimen is loaded until crack propagation begins at load, P_1. As the crack extends to the next mark on the specimen, the load drops to P_2. The specimen is then unloaded and the deflection returned to zero. The test is then repeated to the next increment of crack extension. Five increments of loading, from 38.1 mm (1.5 in.) to 102 mm (4 in.), are used and an average value of G_{Ic} determined from the relationship

$$G_{Ic} = \frac{1}{2b\Delta c} (P_2\delta_2 - P_2\delta_1) \qquad (3)$$

where δ_1 and δ_2 are deflections associated with loads P_1 and P_2, respectively; b is the specimen width; and Δc is the increment of crack extension. A typical loading and unloading curve is shown in Fig. 11. Equation 3 is based on the area between the loading and unloading curves, that is, the energy lost during incremental crack extension for a material undergoing elastic response. This approach also assumes that all of the energy change is absorbed in moving the delamination, that is, no other damage is induced. Other data reduction schemes have been utilized in conjunction with the double-cantilever-beam (DCB) method and are discussed in detail in Ref 6. Materials considered in conjunction with current work displayed linear elastic behavior in the DCB test. Thus, Eq 3 was utilized for all data reduction. For materials displaying significant time-dependent or inelastic behavior or both, the area method would have to be modified.

Typical results from the DCB test are shown in Table 1. In addition to the

TABLE 1—Mode I delamination, G_{Ic} 10^{-2} J/cm^2 (lb/in.²).

Material	V_f	DCB	90-deg CN	FED
AS-1/AF-R-E350	0.70	1.31 (0.752)	2.31 (1.32)	1.94 (1.11)
AS-1/3502	0.70	1.40 (0.801)	1.54 (0.881)	2.67 (1.53)
AS-1/polysulfone	0.54	6.55 (3.74)	...	...
AS-1/ATQ	0.55	3.68 (2.10)	...	0.298 (0.170)
AS-1/3502 [90₂/0₈/90₂]ₛ	0.70	2.80 (1.60)	...	...
AS-1/3502 AF-163U adhesive inlay	0.70	12.8 (7.28)	...	...

two resins displayed in Fig. 1, data are presented for composites containing Hercules' 3502 resin system and an Air Force developed acetylene terminated quinoxaline (ATQ) resin system. With the exception of the 0/90 laminate, all of the results are for 24-ply 0-deg composites.

Note that considerable improvement in energy release rate is obtained with the ductile polysulfone resin system. In making comparison with the brittle resin systems, however, the difference in fiber volume content should be considered.

Whenever data from a DCB test is interpreted, the failure mode and failure surfaces must be carefully scrutinized. For example, the AS-1/ATQ composite yields a relatively high value of G_{Ic} when compared to the epoxy resin composites. The failure surface reveals, however, a brittle failure with many bare fibers exposed. Thus, the crack did not stay along the centerline of the laminate. Some of the energy was expended in creating new surfaces around the fibers. In other matrix-dominated tests, such as the edge-delamination coupon that will be discussed later, the resin appears to be very brittle. Similar results are noted in the AS-1/3502 bidirectional laminate where the crack moves to the 0/90 interfaces above and below the centerline. In fact, it has been noted by Nicholls and Gallagher [7] that multiple failure regions with different energy release rates develop within the same specimen for angle-ply laminates. Since 0-deg composites appear to produce the lowest energy release rate, they appear to provide the most useful data from both a structures and materials characterization viewpoint.

It should also be noted from Table 1 that considerable improvement in energy release rate can be obtained by using ductile inlays such as an AF-163U adhesive.

Current results indicate that the DCB test can provide a useful tool for characterizing interlaminar Mode I energy release rate.

90-Deg Center-Notch Coupon

A 90-deg unidirectional tensile coupon with a center crack provides an alternate approach to the DCB test for determining 0-deg Mode I interlaminar strain energy release rate. The in-plane fiber packing geometry does not provide as clear a fracture plane through the resin as does a DCB test performed on a unidirectional laminate. Thus, a 90-deg center-notch (CN) specimen is likely to produce a somewhat higher value of G_{Ic} compared to a 0-deg DCB test.

Results for the 90-deg CN test are shown in Table 1 for AS-1/AF-R-E350 and AS-1/3502 unidirectional composites. These tensile coupons measured 229 mm (9 in.) in length, while the widths of the specimens were varied depending on the crack length. Three crack sizes were utilized, measuring 5.1 mm (0.2 in.), 10.2 mm (0.4 in.), and 15.2 mm (0.6 in.) for widths of 25.4 mm (1 in.), 38.1 mm (1.5 in.), and 51 mm (2 in.), respectively. The center notch was machined in the specimen by first drilling a 7.4-mm (0.29-in.) diameter hole in the center of the specimen and then using a Lastec diamond wire saw having a

1.127-mm (0.005-in.) diameter wire to machine a straight crack parallel to the fibers on each side of the hole. No attempt was made to sharpen the ends of the crack.

All CN data were adjusted for finite width using the relationship [8]

$$\sigma_n^\infty = \sigma_n[1 + 0.128(a/W) - 0.288(a/W)^2 + 1.53(a/W)^3] \qquad (4)$$

where σ_n is the strength of a tensile coupon with finite width $2W$, containing a center crack of length $2a$, and σ_n^∞ is the notch strength of an equivalent plate of infinite extent.

The average stress criterion was utilized for determining a critical stress intensity factor, k_{Ic}. In particular, an inherent flaw model was assumed, that is

$$k_{Ic} = \sigma_n^\infty \sqrt{\pi(a + a_0)} \qquad (5)$$

where a_0 is an inherent flaw. The value of a_0 can be determined from the average stress criterion as described in Ref 1. If σ_0 denotes the unnotched tensile strength and N the total number of notched specimens tested, then

$$a_0 = \frac{1}{N} \sum_{i=1}^{N} \left[\frac{a_i}{\left(\dfrac{\sigma_0}{\sigma_{ni}^\infty}\right)^2 - 1} \right] \qquad (6)$$

where a_i and σ_{ni} are the half crack length and notched strength, respectively, of the ith coupon. In the average stress criterion, it is assumed that failure occurs when the average normal stress, parallel to the load, over some characteristic length adjacent to the crack, equals the unnotched strength of the material. This characteristic dimension is assumed to be independent of crack length. A critical strain energy release rate can be calculated from the relationship [9]

$$G_{Ic} = k_{Ic}^2 \sqrt{\left(\frac{1}{2E_L E_T}\right)\left[\sqrt{\frac{E_L}{E_T}} + \frac{E_L}{E_T G_{LT}} (E_T - 2\nu_{LT} G_{LT}) \right]} \qquad (7)$$

where E_L, E_T, ν_{LT}, and G_{LT} are modulus parallel to the fibers, modulus transverse to the fibers, major Poisson's ratio, and longitudinal shear modulus, respectively, of the unidirectional composite. These properties are shown in Table 2 for the composite systems under consideration in this work.

As anticipated, the results in Table 1 show higher values of G_{Ic} as determined from the 90-deg CN coupon relative to DCB data. This test does, however, provide a reasonable alternative to the DCB test for determining 0-deg critical strain energy release rate under Mode I delamination.

TABLE 2—*Unidirectional-ply properties.*

Material	V_f	E_L, GPa (Msi)[a]	E_T, GPa (Msi)	G_{LT}, GPa (Msi)	v_{LT}
AS-1/AF-R-E350	0.70	145 (21.0)	10.0 (1.45)	5.53 (0.800)	0.30
AS-1/3502	0.70	145 (21.0)	11.4 (1.65)	4.84 (0.700)	0.30
AS-1/ATQ	0.55	105 (15.2)	8.71 (1.26)	4.15 (0.600)	0.30
AS-1/polysulfone	0.54	113 (16.3)	7.53 (1.09)	3.87 (0.560)	0.34

[a]Msi = 10^6 psi.

Free-Edge-Delamination Tensile Coupon

In the free-edge-delamination test (FED), a straight-sided, multidirectional, laminated coupon is loaded in tension. The stacking geometry is chosen such that delamination is induced along the straight-sided free edges. This test was originally proposed by Pagano and Pipes [10] as an interlaminar strength test. O'Brien [11] extended this test to a fracture resistance method by performing an approximate energy release rate analysis of free-edge-delamination coupons.

Laminates of the class $[\pm 30/90]_s$ tend to delaminate under tensile load. This delamination is driven primarily by interlaminar normal tension along the free edges at the laminate centerline. After delamination is initiated along the entire length of the tensile coupon, the crack front moves toward the center of the specimen. Thus, the crack propagation simulates Mode I delamination in a 0-deg unidirectional composite. Using an analysis similar to the one presented by O'Brien [11], an energy release rate relationship can be derived.

$$G_{Ic} = h\epsilon_c^2(1 - E_1^*/E_1)\bar{E}_1 \tag{8}$$

where h is the half laminate thickness (see Fig. 12), ϵ_c is the critical axial strain at delamination, E_1 is the laminate modulus in the load direction as calculated from laminated plate theory, E_1^* is an effective laminate modulus for a delaminated composite as calculated from laminated plate theory, and $\bar{E}_1$ is the experimentally determined value of E_1. If $\bar{E}_1 = E_1$, Eq 8 reduces to the expression derived by O'Brien [11]. This modification empirically adjusts the analytical

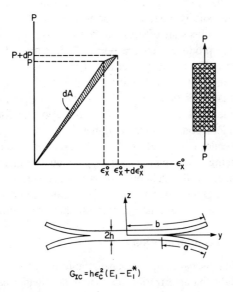

$$G_{IC} = h\epsilon_c^2(E_1 - E_1^*)$$

FIG. 12—*Energy release rate analysis for free-edge-delamination tensile coupon.*

expression for any discrepancy between the theoretical modulus and the actual experimentally determined modulus. In terms of laminated plate stiffnesses [12]

$$E_1 = A_{11} - A_{12}^2/A_{22} \qquad (9)$$

$$E_1^* = \frac{1}{h} \left\{ A_{11} - \left[\frac{A_{12}(A_{12}D_{22} - B_{12}B_{22}) + B_{12}(A_{22}B_{12} - A_{12}B_{22})}{A_{22}D_{22} - B_{22}^2} \right] \right\} \qquad (10)$$

where the laminated plate stiffnesses are expressed in terms of the anisotropic reduced stiffnesses for plane stress, Q_{ij}.

$$(A_{ij}, B_{ij}, D_{ij}) = \int_{-h/2}^{h/2} Q_{ij}(l, z, z^2)dz$$

Equation 10 differs from the expression for E_1^* as derived by O'Brien [11]. Complete derivation of this expression is shown in the Appendix. Note that Eq 10 is derived by considering the upper half of the free-edge coupon as a laminated plate, that is, E_1^* is derived for the stacking geometry [±30/90], which is unsymmetric.

Results from the free-edge delamination test are shown in Table 1. Ply properties used in conjunction with Eqs 9 and 10 are listed in Table 2. The FED test yields somewhat higher values of G_{Ic} than the DCB test for both AS-1/AF-R-E350 and AS-1/3502. The extremely low value of G_{Ic} obtained from the FED test on AS-1/ATQ is likely to be realistic as ATQ is a very brittle resin, yielding only about 0.2% strain to failure in an unnotched 90-deg tension test. As previously discussed, the high value of G_{Ic} obtained from the DCB in conjunction with the ATQ composite is misleading due to the actual failure process.

Because of the complex state of stress in the free-edge zone, the FED specimen does not necessarily produce a pure Mode I delamination. In addition, the assumptions that led to Eq 8 are very crude. The failure is not initiated cleanly along the centerline, as the crack moves to the 90/30 interfaces creating an irregular pattern that is difficult to analyze. The delamination zone is usually very irregular as the crack moves toward the center of the coupon. This makes it difficult to consider a constant crack length as illustrated in Fig. 12. Transverse cracking often precedes delamination. In such cases, all of the energy loss does not go into producing delamination. Finally, for ductile matrix laminates such as AS-1/polysulfone, no edge delamination is produced prior to ultimate laminate failure. Thus, the FED test appears to have some definite drawbacks. This test may have more use as a delamination strength characterization method.

Conclusions and Recommendations

Based on the data discussed in this paper, the following conclusions are made concerning materials characterization for matrix dominated failure modes:

1. Of the current strength tests, 90-deg unidirectional tension appears to be the most useful. In-plane shear can also be informative, but results are often difficult to interpret.

2. A laminate test that interrogates *in situ* transverse tension strength should be added to characterization tests.

3. The short beam shear test appears to be of little value and should be deleted from materials characterization.

4. A 0-deg double-cantilever-beam test appears to be a viable approach for determining Mode I delamination resistance. This method could also be supplemented with a 90-deg center-notch tension test.

5. Despite its drawbacks, a free-edge-delamination tension test performed on laminates of the $[\pm30/90]_s$ class could be used as an alternative to a double-cantilever-beam test for determining Mode I delamination resistance. This method could also be utilized as simply an interlaminar tensile strength test.

The development of a completely sound plan for characterizing matrix-dominated failure modes depends on a thorough understanding of the role of the matrix in composite failure. This need for such an understanding points to areas where additional research is needed. A better understanding of the relationship between matrix stress-strain response and both transverse tension and in-plane shear of unidirectional composites is required. Additional studies on shear as a failure mode, especially under complex states of stress such as in the short beam shear specimen, need to be accomplished. It is not clear, for example, whether a pure shear failure can ever be produced. Although not discussed in this paper, failure processes under compression loading need more clarification before intelligent compression tests can be added to a materials characterization program.

Fracture mechanics appears to offer considerable potential for relating matrix properties to interlaminar fracture resistance. Additional test methods need to be developed or assessed or both, however, in order to be able to completely characterize interlaminar fracture resistance. In particular, viable methods must be made available for measuring Mode II and mixed-mode interlaminar fracture resistance. Since 0-deg delamination appears to provide the weakest failure mode, initial focus of test method development should be on unidirectional composites.

A schematic of potential uses of fracture mechanics, starting with neat resin fracture characterization, is illustrated in Fig. 13. Small compact tension tests can be used to evaluate fiber/matrix interface toughness. In addition to the Mode I tests discussed in this paper, a 0-deg center-notch tension test may also be useful in evaluating composite behavior. For example, matrix ductility or fiber/ matrix interface strength or both may determine whether the crack propagates across the fibers or parallel to the fibers. A simple cantilever beam under bending with a center slit, as proposed by Vanderkley [13], is a potential specimen for measuring pure Mode II. An off-axis unidirectional center-notch specimen, previously utilized by Wu [14], provides one possible approach to mixed-mode

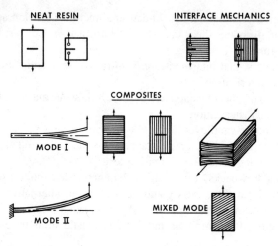

FIG. 13—*Some potential fracture mechanics tests for characterizing matrix-dominated composite properties.*

fracture. In any fracture mechanics characterization, at least two different test methods should be utilized to measure a particular fracture mode. This will help to ascertain as to whether the property is independent of test method and specimen geometry, and is indeed a material property.

Acknowledgment

This work is part of a continuing in-house research program on advanced composites sponsored by the Materials Laboratory of the Wright Aeronautical Laboratories.

The authors wish to acknowledge R. Cornwell, W. Ragland, J. Camping, and R. Esterline of the University of Dayton Research Institute for specimen fabrication and testing.

APPENDIX

Strain Energy Release Rate Analysis for Free-Edge-Delamination Tensile Coupon

Consider the free-edge-delamination tensile coupon shown in Fig. 12. The strain energy release rate due to delamination can be determined from a change in compliance as displayed by the load-strain curve in Fig. 12. For a specimen containing an edge delamination of length a, crack extension from a to $a + da$ induces a change in compliance that results in a loss of strain energy. The strain energy lost for a linear elastic body is simply the area, dA, between the loading and unloading curves. Thus

$$G_1 = \frac{1}{4} \left(P \frac{d\epsilon_x^0}{da} - \epsilon_x^0 \frac{dP}{da} \right) \tag{11}$$

The desired relationship for G_1 can be determined by considering the upper half of the tensile coupon in Fig. 12 over the interval $0 \leq y \leq b$ as a laminated plate. Then

$$\int_0^b N_x \, dy = \frac{P}{4} \tag{12}$$

where N_x is the force resultant per unit length of the plate. If E_1 denotes the axial modulus of the laminate prior to delamination, then

$$\int_0^{(b-a)} N_x \, dy = hE_1(b-a)\epsilon_x^0 \tag{13}$$

where ϵ_x^0 is the mid-plane strain of the laminate in the axial direction. The axial modulus, E_1, is given by Eq 9 that is a well-known relationship derived from laminated plate theory [12].

It now remains to evaluate the integral in Eq 12 over the interval $(b-a) \leq y \leq b$. To do this, it is assumed that the cross section opens up like a double-cantilever beam (see Fig. 12), as suggested by Pagano and Pipes [10], due to the unsymmetric nature of the delaminated portion of the plate. The constitutive relationships for a laminated plate in matrix form are [12]

$$\left[\frac{N}{M}\right] = \left[\begin{array}{c|c} A & B \\ \hline B & D \end{array}\right] \left[\frac{\epsilon^0}{\kappa}\right] \tag{14}$$

where N and M are force and moment resultant vectors, respectively, ϵ^0 is the mid-plane strain vector, and κ denotes plate curvatures. The following assumptions are made

$$N_x = \text{constant}, \ N_y = M_y = 0 \tag{15}$$
$$w = w(y)$$

where w is the plate deflection. The assumption in Eq 15 concerning w leads to κ_y as the only nonvanishing curvature. For laminates of the class under consideration in the free-edge-delamination tension test, the in-plane properties are orthotropic ($A_{16} = A_{26} = 0$). Using Eq 15 in conjunction with Eq 14, one obtains the following constitutive relationships

$$N_x = A_{11}\epsilon_x^0 + A_{12}\epsilon_y^0 + B_{12}\kappa_y \tag{16}$$
$$N_y = 0 = A_{12}\epsilon_x^0 + A_{22}\epsilon_y^0 + B_{22}\kappa_y \tag{17}$$
$$M_y = 0 = B_{12}\epsilon_x^0 + B_{22}\epsilon_y^0 + D_{22}\kappa_y \tag{18}$$

Solving Eqs 17 and 18 for ϵ_y^0 and κ_y and substituting the results into Eq 16, one obtains the following result

$$\int_{(b-a)}^a N_x \, dy = h E_1^* a \epsilon_x^0 \tag{19}$$

where E_1^* is defined by Eq 10. The desired relationship between P and ϵ_x^0 is now obtained by combining Eqs 12, 13, and 19 with the result

$$\left[E_1 - \frac{a}{b}(E_1 - E_1^*)\right]\epsilon_x^0 = \frac{P}{4bh} \tag{20}$$

Substituting Eq 20 into Eq 11, one obtains

$$G_l = h \, \epsilon_x^2 \, (E_1 - E^*) \tag{21}$$

Note that in this analysis, ϵ_x^0 is assumed to be constant across the entire width of the plate, while N_x is constant, but of different magnitude, in the regions $0 \leq y \, (b\text{-}a)$ and $(b\text{-}a) \leq y \leq b$. This is a result of the clamped end conditions in the tensile coupon.

References

[1] Whitney, J. M., Daniel, I. M., and Pipes, R. B., *Experimental Mechanics of Fiber Reinforced Composite Materials*, Society for Experimental Stress Analysis, Brookfield Center, Conn., 1982, pp. 192–196.

[2] Chamis, C. C., "Micromechanics Strength Theories," *Composite Materials, Vol. 5, Fracture and Fatigue*, L. J. Broutman, Ed., 1974, pp. 94–153.

[3] Ashton, J. E., Halpin, J. C., and Petit, P. H., *Primer on Composite Materials: Analysis*, Technomic Publishing Co., Stanford, Conn., 1969.

[4] Browning, C. E., Abrams, F. L., and Whitney, J. M., "A Four-Point Shear Test for Graphite/ Epoxy Composites," *Composite Materials: Quality Assurance and Processing, ASTM STP 797*, American Society for Testing and Materials, Philadelphia, 1983.

[5] Flaggs, D. L. and Kural, M. H., *Journal of Composite Materials*, Vol. 16, 1982, pp. 103–116.

[6] Whitney, J. M., Browning, C. E., and Hoogsteden, W., *Journal of Reinforced Plastics and Composites*, Vol. 1, 1982, pp. 297–313.

[7] Nicholls, D. J. and Gallagher, J., *Journal of Reinforced Plastics and Composites*, Vol. 2, 1984, pp. 2, 17.

[8] Brown, W. F., Jr., and Srawley, J. E. in *Plane Strain Crack Toughness Testing of High Strength Metallic Materials, ASTM STP 410*, American Society for Testing and Materials, Philadelphia, 1966, p. 11.

[9] Paris, P. C. and Sih, G. C. in *Fracture Toughness Testing, ASTM STP 381*, American Society for Testing and Materials, Philadelphia, 1965, p. 30.

[10] Pagano, N. J. and Pipes, R. B., *International Journal of Mechanical Sciences*, Vol. 15, 1973, pp. 679–684.

[11] O'Brien, T. K. in *Damage in Composite Materials, ASTM STP 775*, K. L. Reifsnider, Ed., American Society for Testing and Materials, Philadelphia, 1982, pp. 140–167.

[12] Jones, R. M., *Mechanics of Composite Materials*, McGraw-Hill, New York, 1975, pp. 147–155.

[13] Vanderkley, P. S., "Mode I–Mode II Delamination Fracture Toughness of a Unidirectional Graphite/Epoxy Composite," Texas A & M University Report No. MM3724-81-15, Dec. 1981.

[14] Wu, E. M., *Journal of Applied Mechanics*, Vol. 34, 1967. pp. 967–974.

T. Kevin O'Brien[1]

Mixed-Mode Strain-Energy-Release Rate Effects on Edge Delamination of Composites

REFERENCE: O'Brien, T. K., "**Mixed-Mode Strain-Energy-Release Rate Effects on Edge Delamination of Composites**," *Effects of Defects in Composite Materials, ASTM STP 836*, American Society for Testing and Materials, 1984, pp. 125–142.

ABSTRACT: Unnotched $[\pm\theta/0/90]_s$ graphite/epoxy laminates, designed to delaminate at the edges under static and cyclic tensile loads, were tested and analyzed. The specimen stacking sequences were chosen so that the total strain-energy-release rate, G, for edge delamination was identical for all three layups. However, each layup had different percentages of crack-opening and shear-mode strain-energy-release rates, G_I and G_{II}, respectively. Results with composites made from T300 graphite fibers and 5208 epoxy, a brittle resin, indicated that only G_I contributed to delamination onset under static loading. However, results with composites made from C6000 fibers and H205 epoxy, a tougher resin, indicated that the total G governed the onset of edge delaminations under cyclic loads. In addition, for both materials, the threshold level of G for delamination onset in fatigue was significantly less than the critical G_c measured in static tests. Furthermore, although the C6000/H205 material had a much higher static G_c than T300/5208, its fatigue resistance was only slightly better. A series of mixed-mode tests, like the ones in this study, may be needed to evaluate toughened-resin composites developed for highly strained composite structures subjected to cyclic loads.

KEY WORDS: composite materials, graphite/epoxy, delamination, mixed-mode, fracture mechanics, toughness, fatigue (materials)

Nomenclature

a	Delamination size
Δa	Incremental delamination size
$E_{[90]}$	Stiffness of a 90-deg ply in axial (loading) direction
$E_{[\pm45/0]}$	Axial stiffness of a laminate containing +45, −45, and 0-deg plies
E_{LAM}	Axial laminate stiffness calculated from laminated plate theory
E^*	Axial stiffness of a laminate completely delaminated along one or more interfaces

[1] Research scientist, Structures Laboratory, U.S. Army Research and Technology Laboratories (AVRADCOM), NASA Langley Research Center, Hampton, Va. 23665.

G Strain-energy-release rate associated with edge delamination
G_c Critical strain-energy-release rate for delamination onset
G_I, G_{II}, G_{III} Strain-energy-release rate components due to opening, interlaminar shear, and out-of-plane shear fracture modes
G_{Ic} Critical value of Mode I strain-energy-release rate for delamination onset
h Ply thickness
N Number of fatigue cycles
n Number of plies
t Laminate thickness
ϵ Nominal tensile strain
ϵ_c Nominal tensile strain at delamination onset
ϵ_{MAX} Maximum cyclic strain level in fatigue
θ Fiber orientation angle in laminate plies

One major obstacle to achieving the full weight-saving potential of advanced composite materials in large, highly strained, primary aircraft structures is the tendency for these materials to delaminate. Delamination often results in loss of stiffness, strength, and fatigue life [1–4].[2] Delamination failure criteria are needed to predict the onset and growth of delaminations. One measure of delamination resistance under static loading is interlaminar fracture toughness. The interlaminar fracture toughness, G_c, of a composite laminate is the critical value of the strain-energy-release rate, G, required to cause a delamination to grow. Previously [1], the critical strain-energy-release rate measured at delamination onset in eleven-ply $[\pm30/\pm30/90/\overline{90}]_s$,[3] NARMCO T300/5208[4] graphite/epoxy laminates was used to predict delamination onset in layups with other thicknesses and stacking sequences. Good agreement was found between measured and predicted delamination onset strains. However, before delamination criteria can be generated with confidence, the relative contributions of interlaminar tension and shear to the formation and growth of delaminations must be identified.

The primary goal of this investigation was to determine the effect of different mixed-mode (interlaminar tension and shear) strain-energy-release rate percentages on the formation of edge delaminations in unnotched laminates subjected to static and cyclic tensile loads. Unnotched graphite/epoxy laminates, designed to delaminate at specific interfaces, were analyzed and tested. A parametric study was performed to optimize laminate layups for G_c measurement (see Appendix). A combination of stacking sequences was chosen so that the total strain-energy-release rates were the same, but the percentages of interlaminar tension, G_I, and

---

[2] The italic numbers in brackets refer to the list of references appended to this paper.
[3] The bar over the center 90-deg ply denotes one half of a ply thickness.
[4] Use of trade names or manufacturers does not constitute an official endorsement, either expressed or implied, by the National Aeronautics and Space Administration or AVRADCOM.

interlaminar shear, G_{II}, were very different. A closed-form equation [1] was used to calculate total strain-energy-release rates. A quasi-three-dimensional finite-element analysis [5] was used to determine G_I and G_{II} percentages using the technique outlined previously [1].

Static tests were performed on T300/5208 laminates. Measured delamination onset strains and corresponding strain-energy-release rates were compared for the layups with the same total G but with different G_I and G_{II} percentages. Results were also compared to G_{Ic} measurements from double cantilever beam tests.

Static and fatigue tests were performed on Hexcel C6000/H205 graphite/epoxy laminates. The number of cycles to delamination onset at prescribed maximum cyclic strains (and corresponding strain-energy-release rates) were compared for two layups with the same total G but again with different G_I and G_{II} percentages.

The secondary goal of this investigation was to determine if these static and fatigue tests could be used to establish the relative delamination resistance of different composite materials. Therefore, static and fatigue results for T300/5208 and C6000/H205 were compared.

Static Behavior

Delamination Onset Prediction

Previously, the onset of 0/90 interface edge delaminations in $[+45_n/-45_n/0_n/90_n]_s$ $(n = 1,2,3)$ T300/5208 graphite/epoxy laminates were predicted from a closed-form equation for the total strain-energy-release rate, G, for edge delamination [1]. This equation

$$G = \frac{\epsilon^2 t}{2} (E_{LAM} - E^*) \qquad (1)$$

is independent of delamination size. The stiffness quantities, E_{LAM} and E^*, were calculated from laminated plate theory and the rule of mixtures [1,2]. To predict delamination onset, the critical value, G_c, was calculated from Eq 1 using the critical strain, ϵ_c, measured at the onset of $-30/90$ interface edge delamination in tension tests on $[\pm30/\pm30/90/\overline{90}]_s$ laminates. Then, Eq 1 was rearranged, and this G_c was used to predict ϵ_c in the $[+45_n/-45_n/0_n/90_n]_s$ laminates. The predictions [1] agreed closely with the data reported [6]. In this study, additional 8-ply and 16-ply specimens were tested to verify these results. Figure 1 shows that the predicted and newly measured values of ϵ_c also agreed well. The arrow in the stacking sequence shown in Fig. 1 indicates that delaminations were modeled in the 0/90 interfaces. Figure 2 shows a micrograph of the specimen edge showing the delamination location through the thickness for the 8-ply laminates. Delaminations formed at, and wandered between, the 0/90 interfaces

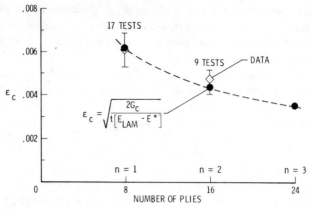

FIG. 1—*Delamination onset prediction, $[+45_n/-45_n/0_n/90_n]_s$ graphite epoxy.*

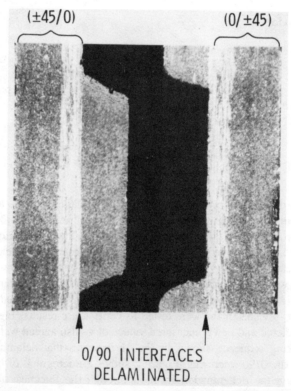

FIG. 2—*Delaminated straight edge, $[\pm45/0/90]_s$ T300/5208 laminate.*

where they were assumed to occur. This same detail was observed for the 16-ply laminates.

Because these results verify the thickness dependence of delamination onset observed in the data and show good quantitative agreement with predicted delamination onset strains, the strain-energy-release rate appears to be a viable parameter for predicting delamination behavior. However, G, as calculated using Eq 1, is a total strain-energy-release rate consisting of both G_I and G_{II}. As previously mentioned, to evaluate the relative contributions of the two fracture modes to edge delamination, three different quasi-isotropic layups, each with the same total G but different percentages of G_I and G_{II}, were analyzed and tested. The total G was calculated from Eq 1, and the relative percentages of G_I and G_{II} were calculated by a finite-element analysis [5] using the technique outlined previously [1].

Analysis

Three different quasi-isotropic layups were analyzed. Laminate A was $[\pm 45/0/90]_s$, Laminate B was $[0/\pm 45/90]_s$, and Laminate C was $[45/0/-45/90]_s$. All three have 90-deg plies in the center to create high interlaminar tensile stresses at the $\theta/90$ interfaces, where θ is either 0 or -45 deg. Delaminations were modeled at the $0/90$ interfaces for Laminate A, and at the $-45/90$ interfaces for Laminates B and C.

Because these layups are all quasi-isotropic, their initial laminate stiffnesses, E_{LAM}, are identical. The rule of mixtures equation for stiffness after the $\theta/90$ interfaces are completely delaminated is

$$E^* = \frac{6E_{[\pm 45/0]} + 2E_{[90]}}{8} \tag{2}$$

The stiffness of a $[\pm 45/0]_s$ laminate, calculated from laminated plate theory, is identical for all three permutations of $+45$, -45, and 0-deg plies in Layups A, B, and C. Hence, E^* is also identical for the three layups. Therefore, the total strain-energy-release rate calculated from Eq 1, for a given nominal strain, ϵ, and laminate thickness, t, will be identical for Laminates A, B, and C.

Figure 3 shows the finite-element meshes used to calculate strain-energy-release rates. Delaminations were modeled in the $\theta/90$ interfaces previously specified. The virtual crack extension technique outlined in Ref *1* was used to compute strain-energy-release rates from nodal forces and displacements calculated before and after an incremental delamination extension, respectively. Figure 4 shows G calculated with the coarse finite-element mesh and a remote strain of 0.004, plotted as a function of delamination size normalized by ply thickness. As was previously observed for other layups [7–10], G_I, and total G are independent of delamination size, once the delamination has grown one to two ply thicknesses in from the specimen edge. As indicated in Fig. 4, the

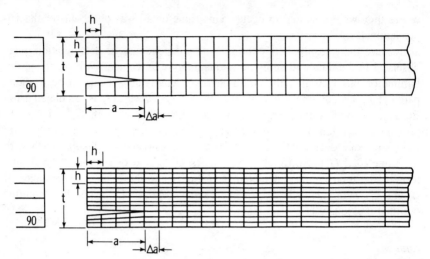

FIG. 3—*Finite-element grids for calculating strain-energy-release rates;* (top) *coarse mesh and* (bottom) *fine mesh.*

total G calculated from finite-element analysis by summing G_I and G_{II} agrees closely with the total G calculated independently from Eq 1. Figure 4 also shows the percentage of the total G attributed to the opening mode, G_I. The opening mode, G_I, was 85%, 57%, and 28% of the total G for Laminates A, C, and B, respectively. The interlaminar shear mode, G_{II}, constituted the remainder. The out-of-plane shear mode, G_{III}, was negligible for all three layups.

Refinement of the coarse finite-element mesh from one to three elements through a ply thickness had a negligibly small effect on the G_I percentages, as

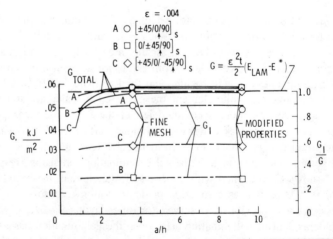

FIG. 4—*Strain-energy-release rates from finite-element analysis.*

shown in Fig. 4 for $a/h = 3.7$. The G curves in Fig. 4 were generated assuming material properties in Table 1 [6]. In addition, a set of three-dimensional material properties were used, based on the work of Kriz [11], to account for the different in-plane and through thickness moduli and Poisson's ratios. These modified properties had only a small influence on the G_I percentages, as shown in Fig. 4 for $a/h = 9.3$.

Experiments

Two 305 by 305-mm panels were made for each of the three layups using the same roll of T300/5208 graphite/epoxy prepreg. Each panel was cured in an autoclave using the manufacturer's prescribed curing cycle. Five 254 by 38-mm (10 by 1.5-in.) coupons were then cut from each panel and tested in tension using apparatus and procedures described in Ref *1* and *2*. A pair of linear variable differential transducers (LVDTs) were mounted on either side of the specimens to measure nominal strain over a 102-mm (4-in.) gage length. Dye-penetrant-enhanced radiographs were taken to confirm the onset of edge delamination, which was indicated by a deviation from a linear stress-strain curve. Photomicrographs of specimen edges shown in Figs. 2 and 5 indicated that delaminations formed only in the interfaces that were assumed to delaminate in the analysis.

Results and Discussion

Laminates A, B, and C all have the same total G. Therefore, if total G_c governs the onset of edge delamination, ϵ_c would be identical for all three layups. However, if only G_I governs delamination onset, then ϵ_c would be lowest for the layup with the highest G_I percentage. Hence, ϵ_c would be lowest for A, highest for B, with C somewhere in between. Figure 6 shows the measured ϵ_c values. The symbols represent the mean of ten tests, and the brackets show the scatter. The mean values of ϵ_c were lowest for Layup A and highest for Layup B, with C in between. Hence, G_I, and not the total G, appears to control the onset of edge delamination for static loading.

Figure 7 shows the critical values of strain-energy-release rate, G_c, calculated from Eq 1 with the ϵ_c data for the three layups. The open symbols and brackets represent the mean values and scatter bands, respectively. The solid symbols represent corresponding values of G_{Ic} calculated from the percentage of G_c that was due to G_I. For Laminates A and C, these G_{Ic} values agree fairly well with G_{Ic} data from unidirectional double cantilever beam (DCB) tests [12]. However,

TABLE 1—*Lamina properties.*

	E_{11}, GPa	E_{22}, GPa	G_{12}, GPa	ν_{12}
T300/5208	134	10.5	5.5	0.30
C6000/H205	124	8.4	5.3	0.33

(+45/0/-45)   (-45/0/45)   (0/+45/-45)   (-45/+45/0)

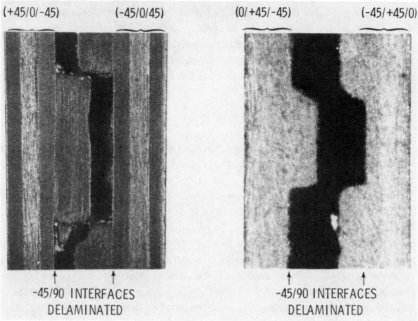

↑   ↑   ↑   ↑
-45/90 INTERFACES   -45/90 INTERFACES
DELAMINATED   DELAMINATED

FIG. 5—*Delaminated straight edges;* (left) $[+45/0/-45/90]_s$ *and* (right) $[0/\pm45/90]_s$ *T300/ 5208 laminates.*

for Laminate B, which had the lowest G_I percentage, the estimated G_{Ic} was well below G_{Ic} measured by the DCB tests.

Both Laminates B and C exhibited large scatter in G_c data and, hence, in G_{Ic} values. Figure 8 shows that both of these laminates developed many 90-deg ply cracks before the onset of delamination, as seen in sequential dye-penetrant-

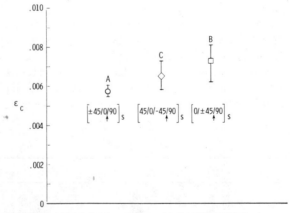

FIG. 6—*Delamination onset strains for T300/5208 laminates.*

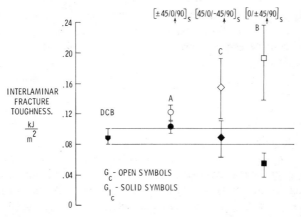

FIG. 7—*Comparison of interlaminar fracture toughness measurements for T300/5208.*

enhanced radiographs taken at load increments throughout the tests. Previously, Talug and Reifsnider [13,14], and more recently Crossman, Wang, and Law [8,9], showed that interlaminar tensile stresses can develop at ply interfaces at matrix crack tips. Perhaps, the interaction of these stresses with the interlaminar edge stresses due to Poisson's mismatch led to the large G_c scatter observed in Layups B and C that developed many cracks before delamination onset. Furthermore, in layups like B where the G_I percentage is low, these stresses may have a significant effect on the apparent mean values of G_{Ic}. Hence, care should be taken to avoid extensive 90-deg ply cracking before delamination onset in layups where the edge delamination test is used to measure interlaminar fracture toughness. Matrix ply cracking can be reduced by optimizing specimen layups to minimize the ϵ_c required to measure a given G_c. A $[\pm35/0/90]_s$ family of layups appears to be optimal (see Appendix). Layups from this family were used to study delamination resistance in fatigue in the next section. In addition, concern

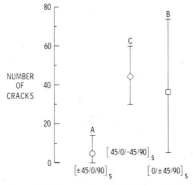

FIG. 8—*Number of 90-deg cracks before delamination, T300/5208.*

about 90-deg ply cracking influencing G_c measurements is diminished for tougher resin composites where 90-deg ply cracking is suppressed [7]. For this reason, C6000/H205 graphite/epoxy composites were analyzed and tested to study mixed-mode effects in fatigue.

Fatigue Behavior

Analysis

Cyclic loading may cause extensive delamination in graphite/epoxy laminates, even for stacking sequences that do not delaminate under static loads. Therefore, it is necessary to characterize delamination resistance in fatigue as well as in static loading. To this end, two layups with the same total strain-energy-release rate, but relatively high and low G_I percentages, were analyzed and tested. Specifically, these layups were $[\pm35/0/90]_s$ and $[0/\pm35/90]_s$. The 35-deg angle was chosen from a parametric study to optimize the layup for the edge delamination test for measuring interlaminar fracture toughness (see Appendix). Both laminates have 90-deg plies in the center to create high interlaminar tensile stresses.

Figure 9 compares G and G_I, calculated at a remote strain of 0.004, for two $[\pm\theta/0/90]_s$ families. Results for both $[\pm\theta/0/90]_s$ families were calculated using the T300/5208 material properties in Table 1 [6]. The quasi-isotropic results were replotted from Fig. 4. As indicated by the arrows in Fig. 9, delaminations were modeled in the 0/90 and $-35/90$ interfaces for the $[\pm35/0/90]_s$ and $[0/\pm35/90]_s$ layups, respectively. The open symbols show the total strain-energy-release rates, calculated by summing G_I and G_{II}. The solid symbol shows that these total G's agree well with values calculated independently from Eq 1. Also shown in Fig. 9 are the G_I calculations from finite-element analyses.

Figure 9 illustrates that changing θ from 45-deg to 35-deg will result in a higher total G at the same remote strain, and a wider range of Mode I percentages for the three different stacking sequence permutations. Furthermore, as was

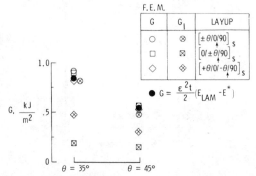

FIG. 9—*Strain-energy-release rates for* $[\pm\theta/0/90]_s$ *laminate families.*

previously noted [7], G_I percentages are controlled by the stacking sequence and are very insensitive to large changes in the matrix-dominated lamina properties, as long as the fiber-dominated moduli are relatively unchanged. Table 2 shows the influence of material properties on G_I percentages for the $[\pm35/0/90]_s$ family of layups. The range of Mode I percentages was slightly greater for T300/5208 than C6000/H205, due primarily to the lower E_{11} value for the C6000/H205 composite (Table 1).

Experiments

Static and constant-amplitude, strain-controlled ($R = 0.2, f = 10$ Hz) cyclic tests were performed on $[\pm35/0/90]_s$ and $[0/\pm35/90]_s$ layups made of C6000/H205 graphite/epoxy. The H205 epoxy was chosen because, as a composite matrix, it has significantly greater interlaminar fracture toughness than the 5208 epoxy [7,15]. Furthermore, 90-deg ply cracking was greatly suppressed in C6000/H205 $[\pm30/\pm30/90/\overline{90}]_s$ laminates used to measure interlaminar fracture toughness [7]. Hence, the tougher H205 matrix was expected to suppress 90-deg ply cracking before delamination in the $[0/\pm35/90]_s$ layup that has a low G_I percentage.

Tests were run until the first indication of delamination onset. Indications involved a combination of visual detection, audible detection, and measured stiffness loss indicated by a discontinuous jump in the load deflection plot during static tests or a drop in cyclic load during the strain-controlled fatigue tests. At the first sign of delamination, loading was stopped and a dye-penetrant-enhanced radiograph was taken to verify the presence of delamination.

Figure 10 shows typical dye-penetrant-enhanced radiographs for the C6000/H205 laminates taken just after delamination onset in fatigue. Figue 11 shows that delaminations formed in the 0/90 interfaces of $[\pm35/0/90]_s$ and in the $-35/90$ interfaces of the $[0/\pm35/90]_s$ layups as modeled. The delaminations wandered through the 90-deg plies to the symmetric θ/90 interfaces. There was no evidence of significant 90-deg ply cracking before delamination during either the static or cyclic loading.

Results and Discussion

Figure 12 shows the delamination onset strains as a function of fatigue cycles for the two C6000/H205 layups chosen from the $[\pm35/0/90]_s$ family. Delamination onset strains for static loading are shown at $N = 0$ in this figure. De-

TABLE 2—*Influence of material properties on* G_I *percentage.*

	$[\pm35/0/90]_s$	$[+35/0/-35/90]_s$	$[0/\pm35/90]_s$
T300/5208	90%	58%	22%
C6000/H205	88%	59%	25%

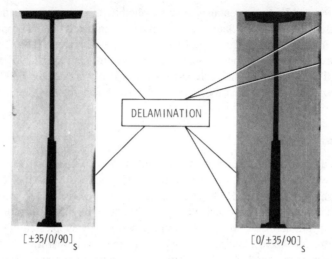

$[\pm 35/0/90]_s$ $[0/\pm 35/90]_s$

FIG. 10—*Radiographs showing delamination onset in fatigue, C6000/H205.*

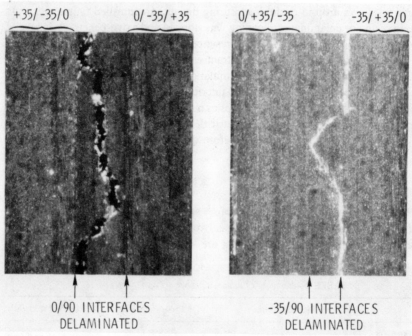

FIG. 11—*Delaminated straight edges after fatigue, C6000/H205,* (left) $[\pm 35/0/90]_s$ *and* (right) $[0/\pm 35/90]_s$.

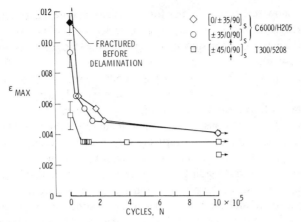

FIG. 12—*Delamination onset as a function of strain and fatigue cycles.*

laminations formed at the edge of [±35/0/90]$_s$ layups before failure. However, the [0/±35/90]$_s$ layups fractured into two pieces before delaminations formed at the edge. These layups have the same total G, but did not delaminate at the same strain. Hence, total G_c does not control delamination onset for static loading. Because the [±35/0/90]$_s$ layup that did delaminate has a very high G_I percentage, then G_I probably plays the dominant role under static loading as was observed earlier for T300/5208. However, quantitative verification would require performing static tests on composites with higher fiber failure strains (see Appendix).

The fatigue data in Fig. 12 show the number of cycles applied at the onset of edge delamination for a range of maximum cyclic strains. Fatigue cycles are plotted on a linear scale to show the steep reduction in the maximum cyclic strain for delamination onset with increasing fatigue cycles. Eventually, a plateau is reached that is tantamount to an endurance limit for delamination onset. Below this level, no delaminations formed. Data points with arrows indicate runouts at 10^6 cycles. Figure 12 demonstrates that delaminations form at different strains in the two layups under static loading. However, the endurance limit strain for delamination onset is identical for the two layups.

Static and constant-amplitude fatigue data for [±45/0/90]$_s$ T300/5208 laminates are also shown in Fig. 12. These fatigue data were generated in load-controlled tests [3]. However, these data can easily be transferred to maximum cyclic strains since no significant stiffness loss occurs before delamination onset [1,2]. The endurance limit for delamination onset, that is, the maximum cyclic strain at which no delamination occurs at 10^6 cycles, is higher for C6000/H205 than for T300/5208. However, because strain endurance limits will decrease with increasing laminate thickness [3] and change with layup [1], comparisons of the fatigue delamination resistance of materials should be presented in terms of strain-energy-release rates.

Figure 13 shows critical G_c values for delamination onset, calculated from Eq 1, as a function of fatigue cycles. For static tests ($N = 0$), G_c values were calculated from the ϵ_c data shown in Fig. 12. For fatigue tests ($N > 0$), G_c values were calculated from the ϵ_{MAX} data shown in Fig. 12. The G_c values for the $[\pm35/0/90]_s$ and $[0/\pm35/90]_s$ layups under static loading are obviously different; however, the threshold G_c values under cyclic loading, calculated from strain endurance limits for the two layups, are identical. Hence, although G_I may govern delamination onset under static loading for the C6000/H205 composite, the total G appears to govern the threshold for delamination onset in fatigue. Obviously, the interlaminar shear that is present may not contribute to delamination under static loading, but contributes fully to delamination under cyclic loading. Of course, the difference in static versus fatigue behavior is not simply a matter of adding the interlaminar shear contribution, because the threshold value of G for fatigue is much less than the static G_c. Furthermore, if we were to consider the H205 matrix material as a toughened resin alternative to 5208, Fig. 13 shows a significant improvement in the static G_c, yet the magnitude of this improvement for the G_c threshold in fatigue is much less. Hence, a series of mixed-mode static and cyclic tests, like those performed in this study, may be needed to evaluate toughened resin composites developed for highly strained composite structures subjected to cyclic loads.

Conclusions

Unnotched composite laminates, designed to delaminate at the edge under static and cyclic tensile loads, were analyzed and tested. The specimen stacking sequences were chosen so that the total strain-energy-release rate, G, for edge

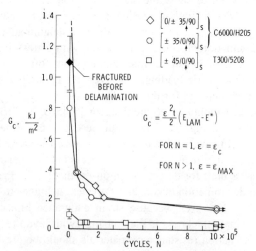

FIG. 13—*Critical G_c as a function of fatigue cycles.*

delamination was identical for all three layups. However, each layup had different percentages of crack opening and shear mode strain-energy-release rates, G_I and G_{II}, respectively. Static and fatigue tests were conducted on T300/5208 and C6000/H205 graphite/epoxy laminates. Based on the analysis and experiments conducted, the following conclusions have been reached.

1. The interlaminar crack opening mode strain-energy-release rate, G_I, controls the onset of edge delamination under static loading.
2. The total mixed-mode (interlaminar tension and shear) strain-energy-release rate controls the onset of edge delamination under cyclic loads.
3. The threshold value of strain-energy-release rate for delamination onset in fatigue is significantly less than the corresponding value measured in a static test.
4. Based on a comparison of T300/5208 and C6000/H205 laminates, the improved delamination resistance provided by the toughened resin (H205) composite was much larger for static loading than fatigue loading.

APPENDIX

Layup Optimization for G_c Measurements

Background

Recently, the edge delamination test was used to measure the interlaminar fracture toughness of toughened resin composites in Ref 7. In that study, eleven-ply $[\pm 30/\pm 30/90/\overline{90}]_s$ laminates, with C6000 graphite fibers reinforcing Hexel 205 and rubber-toughened F185 resins, were tested to determine critical values of mixed-mode interlaminar fracture toughness, G_c. These edge delamination tests were successful in discriminating the improved toughness of the different matrices. However, in order to achieve a better quantitative comparison of interlaminar fracture toughness, layups with a wide range of mixed-mode percentages are needed. In addition, fiber-dominated layups, containing 0-deg plies, may be needed for testing composites with toughened resins (like the rubber-toughened F185 resin) that exhibit material nonlinearity in the individual plies of the $[\pm 30/\pm 30/90/\overline{90}]_s$ layup before delamination onset [7]. Therefore, a parametric study was undertaken to develop optimal layups for the edge delamination test for measuring interlaminar fracture toughness.

The ideal layup for the edge delamination test would be one that requires the fewest number of plies yet delaminates at the lowest possible strain for a given interlaminar fracture toughness. Minimizing the number of plies will help conserve the amount of constituent material needed to make the composite specimen. This is of particular concern when only small quantities of resin are available. Minimizing ϵ_c for measuring a given G_c will help ensure that edge delamination precedes extensive 90-deg ply cracking or fiber failure. Some 90-deg plies must be kept in the center of the laminate to ensure a high Poisson's mismatch [16].

$[\pm \theta/90]_s$ Family Optimization

Perhaps the simplest layup would be a $[\pm \theta/\overline{90}]_s$ five-ply laminate with delaminations occurring in the $-\theta/\overline{90}$ interfaces. As θ varies, the $(E_{LAM} - E^*)$ term in Eq 1 will change. Figure 14 shows the variation in delamination onset strains, ϵ_c, required for various $[\pm\theta/\overline{90}]_s$ laminates to measure an interlaminar fracture toughness of $0.15 \, kJ/m^2$. As shown in Fig. 14, the lowest ϵ_c occurs in the vicinity of 30 deg.

Furthermore, by noting that ϵ_c is proportional to $1/\sqrt{t}$, increasing the laminate thickness, t, will also lower the ϵ_c required to measure a given G_c. This may be accomplished in two ways: first, by increasing the number of angle plies, that is, $[(\pm\theta)_n/\overline{90}]_s$, or second, by increasing the number of 90-deg plies, that is, $[\pm\theta/90_n]_s$, where $n = 1,2,3. \ldots$ Figure 15 shows the effect of both techniques on the ϵ_c required to measure a given G_c. For a given number of 90-deg plies, n, the $[(\pm30)_2/90_n]_s$ layup requires a lower ϵ_c than the $[\pm30/90_n]_s$ layup to measure the same G_c. However, increasing the number of angle plies has the disadvantage of rapidly increasing the laminate thickness and, hence, the amount of material needed to make the specimen. Increasing $\pm\theta$ plies requires adding four new plies each time. As shown in Fig. 15, increasing the number of 90-deg plies will also lower the ϵ_c required to measure a given G_c. However, increasing the number of 90-deg plies has the disadvantage, when taken to extremes, of increasing the contribution of the 90-deg ply cracks to laminate stiffness loss. Such a significant contribution would have to be included in the E^* calculation [1]. In addition, if the number of 90-deg plies grouped in the center is large enough, delaminations can form due to the interlaminar stress fields that develop in the interface at the 90-deg ply crack tips and not due to the large Poisson's mismatch [8,9,17]. However, both these concerns about increasing 90-deg ply thickness are diminished by the tendency for 90-deg ply cracking to be suppressed in composites with toughened-resin matrices [7].

Based on this parametric study, the eleven-ply $[\pm30/\pm30/90/\overline{90}]_s$ layup (shown on Fig. 15 as $[(\pm30)_2/90_n]_s$ where $n = 3/2$) previously used in Ref 7 is a good candidate layup to measure interlaminar fracture toughness. However, because material nonlinearity may appear before delamination for composites with toughened matrix resins [7], an alternate layup was sought for the edge delamination test.

$[\pm\theta/0/90]_s$ Family Optimization

The $[\pm\theta/0/90]_s$ family of layups was considered because including the 0-deg ply, that is, having a fiber-dominated layup, would suppress the material nonlinearity. Figure 14

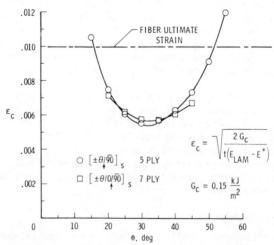

FIG. 14—*Parametric study of variation in delamination onset strains with layup.*

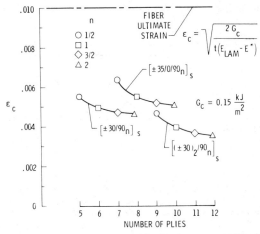

FIG. 15—*Parametric study of variation in delamination onset strains with thickness.*

shows the variation in delamination onset strain, ϵ_c, for various $[\pm\theta/0/90]_s$ layups required to measure an interlaminar fracture toughness of 0.15 kJ/m². The lowest ϵ_c occurs in the vicinity of 35 deg. Figure 15 shows the effect on ϵ_c of increasing 90-deg ply thickness for $[\pm35/0/90_n]_s$ laminates. The same concerns mentioned earlier about increasing 90-deg ply thickness would also apply to this layup, but would be diminished by the tendency for 90-deg ply cracking to be suppressed in toughened resin matrices [7]. As shown in Fig. 15, the largest reduction in ϵ_c results between $n = 1/2$ and $n = 1$. Hence, the eight-ply $[\pm35/0/90]_s$ laminate is a good candidate for the edge delamination test.

Furthermore, the finite-element analysis performed in this study (Table 2) shows that the three permutations of the +35, −35, and 90-deg plies will result in three eight-ply layups with the same total G but a wide range of G_I and G_{II} percentages. Hence, the $[\pm35/0/90]_s$ family of layups should be very useful for evaluating mixed-mode effects in toughened resin composites. One potential disadvantage with these layups, however, is that the range of toughness measurements that can be made is limited by the ultimate tensile strain of the fiber, which controls the nominal failure strain of a fiber-dominated laminate [2]. However, the new high failure strain (> 1.4%) graphite fibers currently on the market will increase the range of G_c measurements possible beyond the 1% limitation illustrated in Figs. 14 and 15.

Summary

As a result of this parametric study, the eleven-ply $[\pm30/\pm30/90/\overline{90}]_s$ and the eight-ply $[\pm35/0/90]_s$ layups appear to be good candidates for the edge delamination test for measuring interlaminar fracture toughness. Furthermore, the three permutations of the eight-ply layup, that is, $[\pm35/0/90]_s$, $[+35/0/-35/90]_s$, and $[0/\pm35/90]_s$, all have the same total G but a wide range of G_I and G_{II} percentages. Hence, these three layups are good candidates for studying mixed-mode effects in toughened resin composites.

References

[1] O'Brien, T. K. in *Damage in Composite Materials, ASTM STP 775*, American Society for Testing and Materials, 1982, pp. 140–167.

[2] O'Brien, T. K. in *Proceedings, SESA/JSME* 1982 Joint Conference on Experimental Mechanics, Hawaii, Part I, Society for Experimental Stress Analysis, Brookfield Center, Conn., May 1982, pp. 236–243.

[3] O'Brien, T. K. in *Fatigue and Creep of Composite Materials,* H. Lilholt and P. Talreja, Eds., Riso National Laboratory, Roskilde, Denmark, 1982, pp. 259–264.

[4] Byers, B. A., "Behavior of Damaged Graphite/Epoxy Laminates Under Compression Loading," NASA CR-159293, National Aeronautics and Space Administration, Aug. 1980.

[5] Raju, I. S. and Crews, J. H., Jr., *Computers and Structures,* Vol. 14, No. 1–2, 1981, pp. 21–28.

[6] Rodini, B. T., Jr., and Eisenmann, J. R. in *Fibrous Composites in Structural Design,* Plenum Publishing Corporation, New York, 1980, pp. 441–457.

[7] O'Brien, T. K., Johnston, N. J., Morris, D. H., and Simonds, R. A., *Journal,* Society for the Advancement of Materials and Process Engineering, Vol. 18, No. 4, July/August 1982, pp. 8–15.

[8] Law, G. E., "Fracture Analysis of $(\pm25/90_n)_s$ Graphite Epoxy Composite Laminates," Ph.D. thesis, Drexel University, Philadelphia, Pa., June 1981.

[9] Crossman, F. W., Warren, W. T., Wang, A. S. D., and Law, G. E., *Journal of Composite Materials Supplement,* Vol. 14, No. 1, 1980, pp. 88–106.

[10] Wang, S. S. in *Proceedings,* 22nd AIAA/ASME/ASCE/AHS Structures, Structural Dynamics and Materials Conference, Part 1, AIAA CP-811, AIAA Paper No. 81-0578, American Institute of Aeronautics and Astronautics, April 1981, pp. 473–484.

[11] Kriz, R. D. and Stinchcomb, W. W., *Experimental Mechanics,* Vol. 19, No. 2, Feb. 1979, pp. 41–49.

[12] Wilkins, D. J., Eisenmann, J. R., Camin, R. A., Margolis, W. S., and Benson, R. A. in *Damage in Composite Materials, ASTM STP 775,* American Society for Testing and Materials, 1982, pp. 168–183.

[13] Talug, A. and Reifsnider, K. L., *Fibre Science and Technology,* Vol. 12, 1979, pp. 201–215.

[14] Reifsnider, K. L. and Talug, A., *International Journal of Fatigue,* Vol. 3, No. 1, Jan. 1980, pp. 3–11.

[15] Bascom, W. D., Bitner, J. L., Moulton, R. J., and Siebert, A. R., *Composites,* Vol. 11, 1980, pp. 9–18.

[16] Pagano, N. J. and Pipes, R. B., *International Journal of Mechanical Science,* Vol. 15, 1973, pp. 679–688.

[17] Crossman, F. W. and Wang, A. S. D. in *Damage in Composite Materials, ASTM STP 775,* American Society for Testing and Materials, 1982, pp. 118–139.

George E. Law[1]

A Mixed-Mode Fracture Analysis of (±25/90$_n$)$_s$ Graphite/Epoxy Composite Laminates

REFERENCE: Law, G. E., "A Mixed-Mode Fracture Analysis of (±25/90$_n$)$_s$ Graphite/Epoxy Composite Laminates," *Effects of Defects in Composite Materials, ASTM STP 836*, American Society for Testing and Materials, 1984, pp. 143–160.

ABSTRACT: An investigation of the fracture mechanisms and behavior of sub-laminate cracking in graphite/epoxy composite laminates is presented. A series of laminates of the form (±25/90$_n$)$_s$ is chosen for analytical modeling and compared with experimental results. By varying the number of 90-deg plies, indicated by n, several important and distinct fracture modes are identified. The thickness of the 90-deg layer is the single parameter that separates these fracture events. The present work focuses in particular on the mechanisms of transverse ply cracking and free-edge delamination.

The energy release rate approach of classical fracture mechanics is applied to describe the crack initiation fracture process. The cracking process is numerically simulated using a finite-element procedure formulated within the framework of ply elasticity and on the assumption of a generalized plane-strain state. The numerical procedure explicitly calculates the Mode I, Mode II, and Mode III components of the strain-energy release rate as a function of crack length. Unit mechanical and unit thermal load conditions are solved, and the results are superposed to obtain the general load case. The analytical model predicts the sequence of occurrence of the fracture modes and the critical onset loads.

The theoretical model is then correlated with published experimental data including tension tests on (±25/90$_n$)$_s$ (n = ½, 1, 2, 3, 4, 6, and 8) laminate coupons manufactured using the T300/934 graphite/epoxy system and with fracture tests using double-cantilever-beam and cracked-lap-shear specimens. Good agreement is obtained between the theoretical and experimental results. The general nature of the present analytical method is readily applicable to composite laminates of more complicated structural geometry or loading conditions or both.

KEY WORDS: composite materials, energy release rate, fracture mechanics, Mode I, Mode II, delamination, transverse crack, graphite/epoxy composites, fatigue (materials)

The fracture behavior of graphite-reinforced-epoxy composites is of current interest, particularly with regard to their durability and damage tolerance. Numerous studies have been performed to determine the critical value of the strain-

[1] Engineering specialist, General Dynamics' Fort Worth Division, Fort Worth, Tex. 76101.

energy release rate [1–5][2] and to examine the interlaminar and intralaminar fracture response of graphite/epoxy composite laminates [6–17].

The purpose of this paper is to predict the initiation loads for transverse cracking and free-edge delamination of the $(\pm 25/90_n)_s$ family of laminates utilizing a mixed Mode I–Mode II fracture criteria. It is shown that the thermal residual stresses due to cool down in the curing process are a significant factor in the fracture analysis.

Experimental Results

The results of an in-depth experimental investigation into the interlaminar and intralaminar fracture behavior of $(\pm 25/90_n)_s$, $n = \frac{1}{2}$, 1, 2, 3, 4, 6, and 8, T300/934 graphite/epoxy test coupons has been previously reported in Ref 7. All test specimens were of the form of straight tensile coupons with a test section 25.4 mm long by 152.4 mm wide (6 in. by 1 in.). The storage and test conditions were maintained at room temperature, 25°C (75°F), and ambient humidity (60% relative humidity) conditions.

As shown schematically in Fig. 1, these laminates may be segregated into three categories based on their fracture behavior: thin ($n = \frac{1}{2}$ and 1), thick ($n = 2$ and 3), and very thick ($n = 4$, 6, and 8) 90-deg layers. The initial fracture of the thin 90-deg layer laminates is a free-edge delamination along the mid-plane of the laminate. The thick 90-deg layer laminates showed transverse ply cracking as their first failure mode, followed by free-edge delamination along the 25/90 interface. Lastly, the very thick 90-deg layer laminates exhibited transverse ply cracking as their first failure mode, followed by a combination of free-edge delamination and transverse-crack-tip delamination. These fracture modes are illustrated in Fig. 2. The experimental onset strain for each fracture event is summarized in Fig. 3. The significance of these tests is that a vast

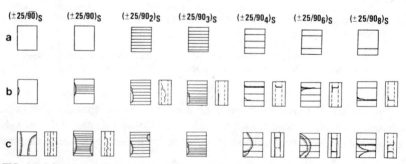

FIG. 1—*Schematic of the fracture sequence in the* $(\pm 25/90_n)_s$ *laminates:* (a) *just prior to edge delamination,* (b) *subsequent to edge delamination, and* (c) *just prior to final failure.*

---

[2] The italic numbers in brackets refer to the list of references appended to this paper.

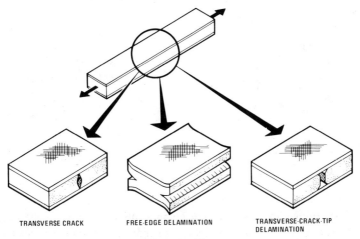

TRANSVERSE CRACK   FREE-EDGE DELAMINATION   TRANSVERSE-CRACK-TIP
                                                      DELAMINATION

FIG. 2—*Schematic of fracture types.*

difference in fracture behavior was observed through simply changing the thickness of the 90-deg ply layers.

Calculation of Energy Release Rate

A numerical technique to calculate the strain-energy release rate using finite-element models has been presented by Rybicki and Kanninen [18]. In the finite-element representation, the continuous stress and displacement fields of the solid are approximated by the nodal forces and displacements, respectively. Figure 4 (*left*) illustrates the finite-element representation of a crack tip region. Here, a crack of Length a is shown with the crack tip at Node f. The finite-element

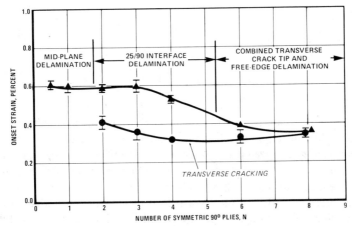

FIG. 3—*Experimental onset strains.*

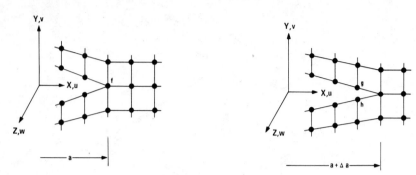

FIG. 4—*Finite-element mesh at crack tip.*

solution determines the displacement components $(u, v, w)_f$, of the crack tip node, f, under a prescribed displacement loading. An incremental crack extension, Δa, is introduced by replacing the crack tip node, f, with two separate nodes, g and h, as shown in Fig. 4 (*right*). With this new crack geometry taken into account, the finite-element solution for the nodal displacements, $(u, v, w)_g$ and $(u, v, w)_h$, are found for Nodes g and h, respectively. The crack extension is then closed by applying equal and opposite forces at Nodes g and h such that their common displacements match the displacements found earlier for Node f.

The work required to close the crack extension is approximated by

$$W = \tfrac{1}{2}[F_x(u_g - u_h) + F_y(v_g - v_h) + F_z(w_g - w_h)] \tag{1}$$

where F_x, F_y, and F_z are the components of the nodal forces required to close Nodes g and h together. Thus, the energy release rate for the three crack extension modes are approximated by

$$G_I = F_y(v_g - v_h)\frac{1}{2\Delta a}$$

$$G_{II} = F_x(u_g - u_h)\frac{1}{2\Delta a} \tag{2}$$

$$G_{III} = F_z(w_g - w_h)\frac{1}{2\Delta a}$$

In the present analysis, the cracking process is modeled using a finite-element procedure formulated within the framework of ply elasticity and based on the assumption of generalized plane strain [8,19]. By judicious selection of the plane of generalized plane strain, the transverse cracking, free-edge delamination, and transverse-crack-tip delamination problems may each be modeled with this finite-element procedure. Both a mechanical load of uniaxial tension and a thermal load due to residual curing stresses are considered. The mechanical and thermal load conditions are solved independently under unit load conditions and are superposed to obtain the combined load condition. The energy release rate as a function of crack length is represented by a set of shape functions as described later. The sets of shape functions used in this study are given in the Appendix. Let f_e and d_e be the crack tip nodal forces and nodal displacements, respectively, due to a unit uniaxial strain, e, and let f_T and d_T be the crack tip nodal forces and nodal displacements due to a unit thermal load, $\Delta T = -1$ deg. Define the mechanical load shape functions for Mode I, Mode II, and Mode III crack growth as

$$C_{EI} = (f_{ey}\, d_{ey}) \frac{1}{2t\Delta a}$$

$$C_{EII} = (f_{ex}\, d_{ex}) \frac{1}{2t\Delta a} \qquad (3)$$

$$C_{EIII} = (f_{ez}\, d_{ez}) \frac{1}{2t\Delta a}$$

where t is the cured lamina ply thickness. The cured ply thickness for the T300/934 material system is 0.132 mm (0.0052 in.). Similarly, define the thermal load shape functions for Mode I, Mode II, and Mode III crack growth as

$$C_{TI} = (f_{Ty}\, d_{Ty}) \frac{1}{2t\Delta a}$$

$$C_{TII} = (f_{Tx}\, d_{Tx}) \frac{1}{2t\Delta a} \qquad (4)$$

$$C_{TIII} = (f_{Tz}\, d_{Tz}) \frac{1}{2t\Delta a}$$

Note that these energy-release-rate shape functions are a function of the non-dimensionalized crack length, a/t. This dependence on crack length is illustrated by the figures in the Appendix. A detailed description of the finite-element techniques used to obtain the shape functions is contained in Ref 8.

The superposition of stress intensity factors for two load conditions is given by [20]

$$K_I = K_{I(1)} + K_{I(2)} \qquad (5)$$

This may be restated in terms of energy release rates as [20]

$$G_I = (\sqrt{G_{I(1)}} + \sqrt{G_{I(2)}})^2 \qquad (6)$$

For the problem at hand, the superposition of the mechanical and thermal loads gives the Mode I, Mode II, and Mode III components of the energy release rate in terms of the unit load shape functions as

$$G_I = (\sqrt{C_{EI}\, e^2\, t} + \sqrt{C_{TI}\, \Delta T^2\, t})^2$$
$$G_{II} = (\sqrt{C_{EII}\, e^2\, t} + \sqrt{C_{TII}\, \Delta T^2\, t})^2 \qquad (7)$$
$$G_{III} = (\sqrt{C_{EIII}\, e^2\, t} + \sqrt{C_{TIII}\, \Delta T^2\, t})^2$$

The Mode III component of the strain-energy release rate was found to be negligible in the ensuing analysis and is only included in the preceding description for completeness.

Fracture Criteria

The condition that leads to fracture for a Mode I crack, such as the transverse cracking and mid-plane delamination problems, occurs when the strain-energy release rate reaches a critical value as

$$G_I = G_{Ic} \qquad (8)$$

where G_{Ic} is the critical strain-energy release rate for Mode I. Combining Eqs 7 and 8 gives the fracture criterion in terms of the unit load shape functions as

$$e_c\, \sqrt{C_{EI}} + \Delta T\, \sqrt{C_{TI}} = \sqrt{G_{Ic}/t} \qquad (9)$$

where e_c refers to the critical failing strain.

The analysis of a combined Mode I–Mode II crack growth, such as the 25/90 interface and transverse-crack-tip delamination problems, requires a mixed-mode failure criterion. The present analysis makes use of a mixed-mode fracture criterion advanced by Wu [21] of the form

$$\sqrt{G_I/G_{Ic}} + G_{II}/G_{IIc} = 1 \qquad (10)$$

where G_{IIc} is the critical strain-energy release rate for Mode II crack growth. Replacing G_I and G_{II} with their shape function representations given by Eq 7 leads to a quadratic equation in e_c. Note that Eq 10 reduces to Eq 8 when $G_{II} = 0$.

Finally, the analysis of a combined transverse-crack-tip delamination (TCTD) and free-edge delamination (FED), as exhibited by the very thick 90-deg layer laminates, requires a three-dimensional solution. In the present analysis, a method is introduced to superpose two two-dimensional solutions. This superposition method is illustrated by Fig. 5. Details of the finite-element modeling techniques are described in Ref 8. Since these two forms of crack growth are orthogonal (the transverse-crack-tip delamination grows in the 0-deg or loading direction and the free-edge delamination grows in the 90-deg direction transverse to the

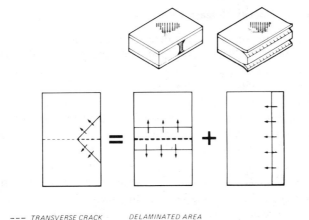

--- *TRANSVERSE CRACK*   *DELAMINATED AREA*

FIG. 5—*Superposition of transverse-crack-tip and free-edge delamination.*

loading), it is proposed that the energy release rates be treated as vector quantities such that

$$G_{TOT} = \sqrt{(G_{TCTD}^2 + G_{FED}^2)} \qquad (11)$$

where G_{TCTD} and G_{FED} refer to the energy release rates for the transverse-crack-tip and free-edge delamination problems, respectively. This "rule" is then applied to each fracture mode individually, and substituted into the previously described mixed-mode fracture criterion given in Eq 10. However, this superposition is not based on rigorous analysis and is thus no replacement for the three-dimensional analysis required to describe this complex problem.

In order to apply the fracture criteria given by Eq 10, it is first necessary to determine the stress-free temperature of the laminate and the initial flaw size. For the T300/934 material system, the laminate curing temperature is 177°C (350°F). Allowing internal stress relaxation and moisture pick-up subsequent to curing, the stress-free temperature will be less than the curing temperature. Experiments conducted by Kim and Hahn [22] on similar composite laminates indicate that the stress-free temperature is approximately 150°C (300°F). In the present analysis, the ambient test condition of 25°C (75°F) and 60% relative humidity is assumed equivalent to a uniform temperature change of $\Delta T = -125°C$ ($-225°F$). The value of the initial flaw size is taken as 0.254 mm (0.010 in.) as determined by an unnotched transverse tension test [8]. Thus, the shape functions are evaluated at a relative crack length of $a/t = 2$. For cases in which the shape functions show a maximum value prior to $a/t = 2$, the maximum value of the shape function is used in the analysis [19].

Critical Strain-Energy Release Rate

Experimental investigations that determine the critical values of the strain-energy release rate for delamination and transverse cracking have only recently become available. Using a double-cantilever-beam specimen, which represents the delamination problem, Vanderkley [1] found values of G_{Ic} as low as 155 J/m^2 (0.887 in-lb/in.^{2}) and Cullen [2] as low as 128 J/m^2 (0.73 in-lb/in.^{2}) for the AS1/3502 material system. Williams [3], utilizing a compact tension specimen to represent transverse cracking, reported values of G_{Ic} equal to 270 J/m^2 (1.54 in-lb/in.^{2}) for the same material system. Williams explained the difference in G_{Ic} values for delamination and transverse cracking based on the morphology of the fracture surface. In the case of delamination (interlaminar fracture), as long as there is no fiber nesting between the plies, the fracture surface is very smooth. For transverse cracking (intralaminar fracture), the fracture surface is not smooth because there is no clear interface for the crack to follow. The fiber involvement in the intralaminar fracture process causes the composite to have a greater resistance (as measured by G_{Ic}) to intralaminar fracture than to interlaminar fracture.

Wilkins [4] performed a series of fracture tests using double-cantilever-beam and cracked-lap-shear specimens made from the AS1/3501-6 material system. Wilkins also found that the morphology of the crack surface significantly affects the fracture toughness. His specimens were carefully prepared to minimize fiber nesting along the fracture surface so that valid values of the critical strain-energy release rate could be measured. He reported a value of G_{Ic} = 140 J/m^2 (0.8 in-lb/in.^{2}) from the double-cantilever-beam tests. Wilkins reported a value of G_{IIc} = 315 J/m^2 (1.8 in-lb/in.^{2}) from his cracked-lap-shear tests that include a combination of Mode I and Mode II. This reported value of G_{IIc} ignores the Mode I contribution that was reported to be 31% of the Mode II component.

For the present analysis of the T300/934 material system, the value of G_{Ic} for transverse cracking is chosen as 270 J/m^2 (1.54 in-lb/in.^{2}) based on Williams's experimental results [3]. The value of G_{Ic} for the delamination problem is taken as 140 J/m^2 (0.8 in-lb/in.^{2}) based on the experimental results of Vanderkley [1], Cullen [2], and Wilkins [4]. The value of G_{IIc} for delamination is chosen as 315 J/m^2 (1.8 in-lb/in.^{2}) based on the experimental data reported by Wilkins [4]. Although this toughness data has not been generated for the T300/934 material system, it is appropriate to apply the basic data of the AS1/3501-6 material system to the T300/934 system because these two epoxies are chemically similar. The AS1/3502 material system has a higher glass transition temperature than the T300/934 system that, in general, translates into a lower fracture toughness.

Results

The result of applying the Mode I and mixed Mode I–Mode II fracture criteria to the $(\pm 25/90_n)_s$ laminates is displayed in Fig. 6. Good agreement is demonstrated between the theory and experiments. Mid-plane, Mode I delamination at

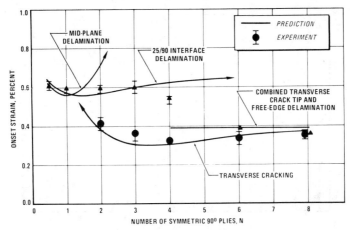

FIG. 6—*Comparison of predicted onset strains and experiment.*

the free edge is the initial predicted fracture mode in the laminate when $n = \frac{1}{2}$ and 1, while transverse cracking in the 90-deg layers is the initial predicted fracture mode for $n = 2$. Free-edge delamination for the laminates of $n = 2$ and 3 is predicted to occur as a mixed Mode I–Mode II fracture at the 25/90 interface subsequent to transverse cracking. Finally, for $n = 4$, 6, and 8, a combined transverse-crack-tip and free-edge delamination along the 25/90 interface is the predicted fracture mode. Each of these predicted fracture modes is as observed in the experiments.

The residual thermal stresses play a significant role in the current fracture analysis. The thermal contribution has been obtained by comparing solutions of Eq 10 for $\Delta T = 0$ and $\Delta T = -125°C\ (-225°F)$. Figure 7 shows the contribution of the residual thermal loading to transverse cracking. For the laminates that exhibit transverse cracking as their first fracture mode ($n = 2$ and above), approximately 50% of the energy released due to transverse cracking as a direct result of the residual thermal load. The contribution of the thermal loading in the delamination analysis is displayed in Fig. 8. Here the thermal load accounts for a minimum of 20% of the fracture energy released. Thus, it is concluded that the thermal residual stresses must be considered in the transverse cracking and delamination analysis.

Conclusions

The mixed Mode I–Mode II fracture analysis predicts the type and nature of the fracture events experimentally observed for the $(\pm 25/90_n)_s$ family of graphite/epoxy composite laminates. The fracture analysis correctly discerns whether transverse cracking of the 90-deg plies or free-edge delamination is the first failure mode. It also predicts the ply interface of free-edge delamination, be it the mid-plane or the 25/90 interface.

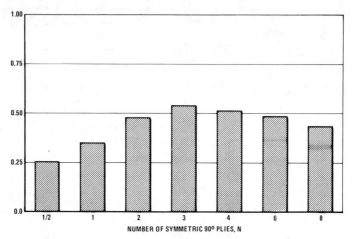

FIG. 7—*Contribution of thermal loads to transverse cracking.*

The mixed Mode I–Mode II fracture analysis also accurately predicts the onset strain for each of the fracture types that were experimentally observed. Excellent agreement was obtained between theory and experiment for transverse cracking. The theory also predicts well the onset strain for free-edge delamination when interaction between transverse cracking and edge delamination is not evident in the form of transverse-crack-tip delamination. The simplistic method presented to superpose the transverse-crack-tip and free-edge delamination analysis produced surprisingly good results when compared with the experiments. However,

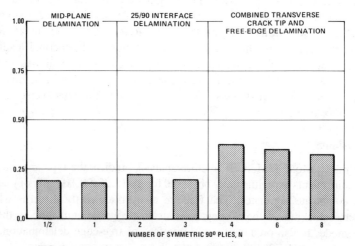

FIG. 8—*Contribution of thermal loads to free-edge delamination.*

this superposition is not based on rigorous analysis and is thus no replacement for the three-dimensional analysis required to describe this complex problem.

The thermal residual loading caused by cool-down after curing was found to have a significant affect in the fracture load predictions. The thermal loading represents approximately a 50% contribution to the transverse cracking problem and approximately a 20% contribution to the free-edge delamination problem. Although it is not clear that the thermal residual loads must be included in all fracture analyses of fiber-reinforced composite materials, it is apparent that the thermal residual loads cannot be ignored in the analysis of transverse cracking and free-edge delamination.

Acknowledgments

This work was supported, in part, by a research grant from the Air Force Office of Scientific Research to Drexel University and Lockheed Palo Alto Research Laboratory, and, in part, by the Lockheed Independent Research Program.

APPENDIX

Strain energy release rate shape functions required to perform the transverse cracking and edge delamination fracture analysis of the $(\pm 25/90_n)_s$, $n = \frac{1}{2}$, 1, 2, 3, 4, 6, and 8, laminates are shown in Figs. 9 through 20.

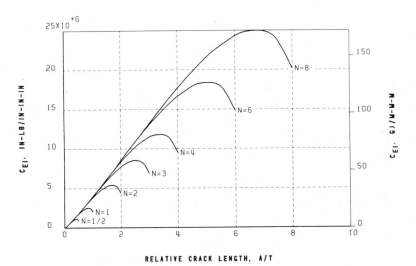

RELATIVE CRACK LENGTH, A/T

FIG. 9—*Mode I mechanical load shape function for transverse cracking.*

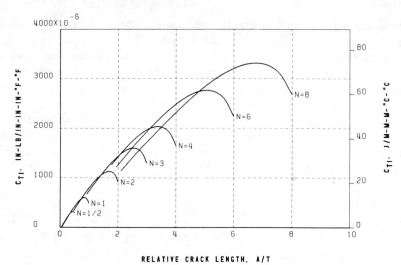

FIG. 10—*Mode I thermal load shape function for transverse cracking.*

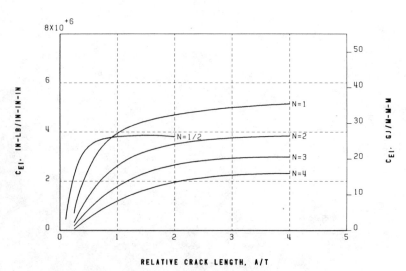

FIG. 11—*Mode I mechanical load shape function for mid-plane delamination.*

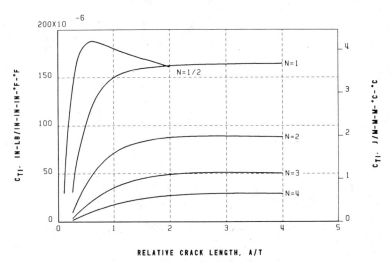

FIG. 12—*Mode I thermal load shape function for mid-plane delamination.*

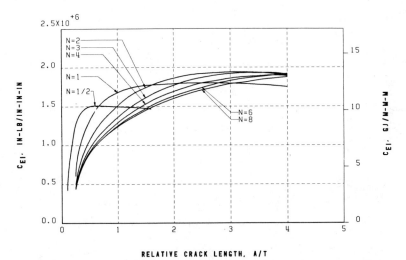

FIG. 13—*Mode I mechanical load shape function for 25/90 interface delamination.*

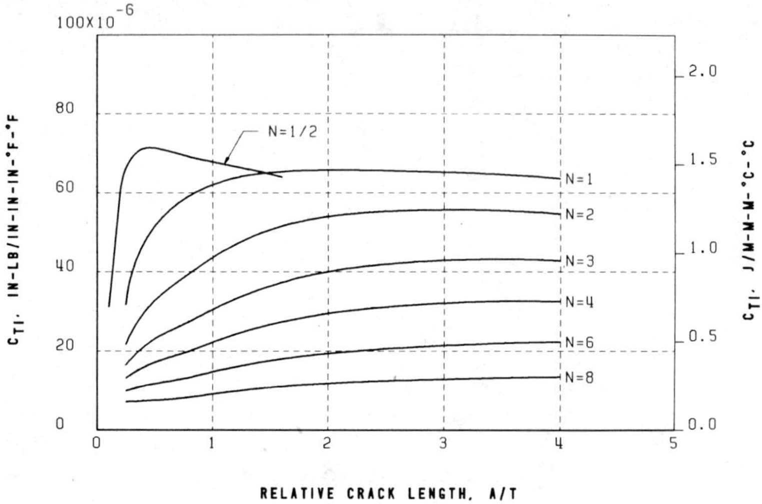

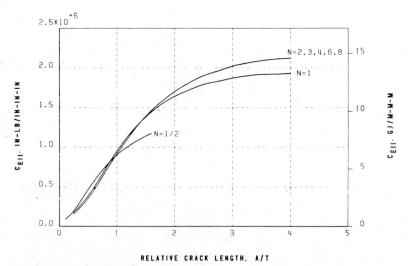

FIG. 14—*Mode I thermal load shape function for 25/90 interface delamination.*

FIG. 15—*Mode II mechanical load shape function for 25/90 interface delamination.*

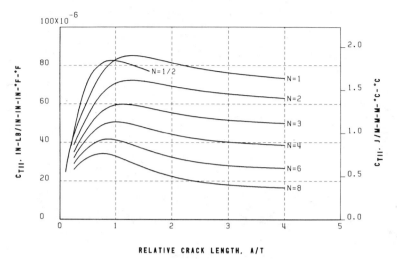

FIG. 16—*Mode II thermal load shape function for 25/90 interface delamination.*

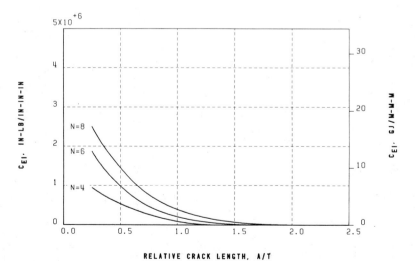

FIG. 17—*Mode I mechanical load shape function for transverse-crack-tip delamination.*

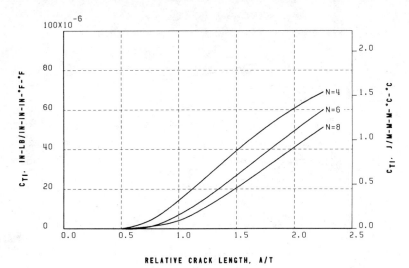

FIG. 18—*Mode I thermal load shape function for transverse-crack-tip delamination.*

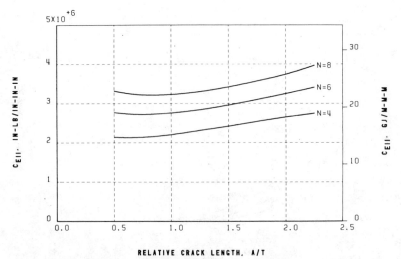

FIG. 19—*Mode II mechanical load shape function for transverse-crack-tip delamination.*

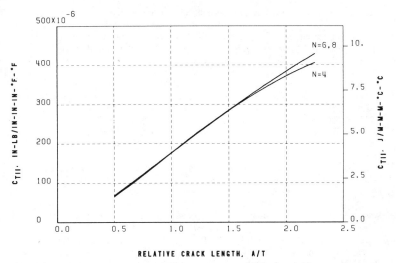

FIG. 20—*Mode II thermal load shape function for transverse-crack-tip delamination.*

References

[1] Vanderkley, P. S., "Mode I–Mode II Delamination Fracture Toughness of a Unidirectional Graphite/Epoxy Composite," Master's thesis, Texas A&M University, 1981.

[2] Cullen, J. S., "Mode I Delamination of Unidirectional Graphite/Epoxy Composite Under Complex Load Histories," Master's thesis, Texas A&M University, 1981.

[3] Williams, D. R., "Mode I Transverse Cracking in an Epoxy and a Graphite Fiber Reinforced Epoxy," Master's thesis, Texas A&M University, 1981.

[4] Wilkins, D. J., "A Comparison of the Delamination and Environmental Resistance of a Graphite-Epoxy and a Graphite Bismaleimide," Final Report, NAV-GD-0037, General Dynamics' Fort Worth Division, 1981.

[5] Wilkins, D. J., Eisenmann, J. R., Camin, R. A., Margolis, W. S., and Benson, R. A., "Characterizing Delamination Growth in Graphite-Epoxy," *Damage in Composite Materials: Basic Mechanisms, Accumulation, Tolerance, and Characterization, ASTM STP 775,* American Society for Testing and Materials, 1982.

[6] Waddoups, M. E., Eisenmann, J. R., and Kaminski, B. E., "Macroscopic Fracture Mechanics of Advanced Composite Materials," *Journal of Composite Materials,* Vol. 5, 1971.

[7] Crossman, F. W., Warren, J., Wang, A. S. D., and Law, G. E., "Initiation and Growth of Transverse Cracks and Edge Delamination in Composite Laminates: Part 2. Experimental Correlation," *Journal of Composite Materials,* Supplemental Vol., 1980.

[8] Law, G. E., "Fracture Analysis of ($\pm25/90_n$)$_s$ Graphite-Epoxy Composite Laminates," PhD thesis, Drexel University, 1981.

[9] Rodini, B. T., Jr., and Eisenmann, J. R., "An Analytical and Experimental Investigation of Edge Delamination in Composite Laminates," *Proceedings,* 4th Conference Fibrous Composites, San Diego, Calif., 1978.

[10] Reifsnider, K. L., Henneke, E. G., and Stinchcomb, W. W., "Delamination in Quasi-Isotropic Graphite-Epoxy Laminates," *Composite Materials: Testing and Design, ASTM STP 617,* American Society for Testing and Materials, 1977.

[11] Reifsnider, K. L. and Masters, J. E., "Investigation of Characteristic Damage States in Composite Laminates," ASME Paper 78-WA/AERO-4, American Society of Mechanical Engineers, 1978.

[12] Bader, M. G., Bailey, J. E., Curtis, P. T., and Parvizi, A., "The Mechanisms of Initiation and Development of Damage in Multi-Axial Fibre-Reinforced Plastics Laminates," *Proceed-*

ings, 3rd International Symposium, Mechanical Behavior of Materials, Cambridge, U.K., Vol. 3, 1979.

[*13*] Parvizi, A., Garrett, K. W., and Bailey, J. E., "Constraint Cracking in Glass Fibre Reinforced Epoxy Cross-Ply Laminates," *Journal Material Sciences,* Vol. 13, 1978.

[*14*] Wang, A. S. D., Law, G. E., and Warren, W. J., "An Energy Method for Multiple Transverse Cracks in Graphite/Epoxy Laminates," *Modern Developments in Composite Materials,* ASME Publication No. G00159, American Society of Mechanical Engineers, 1979.

[*15*] Crossman, F. W. and Wang, A. S. D., "The Dependence of Transverse Cracking and Delamination on Ply Thickness in Graphite/Epoxy Laminates," *Damage in Composite Materials, ASTM STP 775,* American Society for Testing and Materials, 1982.

[*16*] O'Brien, T. K., "Characterization of Delamination Onset and Growth in a Composite Laminate," *Damage in Composites, ASTM STP 775,* American Society for Testing and Materials, 1982.

[*17*] Whitcomb, J. D., "Finite Element Analysis of Instability-Related Delamination Growth," NASA TM-81964, National Aeronautics and Space Administration, 1981.

[*18*] Rybicki, E. F. and Kanninen, M. F., "A Finite Element Calculation of Stress Intensity Factors by a Modified Crack Closure Integral," *Engineering Fracture Mechanics,* Vol. 9, 1977.

[*19*] Wang, A. S. D. and Crossman, F. W., "Some New Results on Edge Effects in Symmetric Composite Laminates," *Journal of Composite Materials,* Vol. 11, 1977.

[*20*] Tada, H., Paris, P. C., and Irwin, G. R., *The Stress Analysis of Cracks Handbook,* Del Research Corporation, Hellertown, Pa., 1973.

[*21*] Wu, E. M., "Crack Extension in Fiberglass Reinforced Plastics," University of Illinois T&AM Report No. 275, 1973.

[*22*] Kim, R. Y. and Hahn, H. T., "Effects of Curing Stress on the First Ply Failure in Composite Laminates," *Journal of Composite Materials,* Vol. 13, 1979.

*Sailendra N. Chatterjee,[1] R. Byron Pipes,[2] and
Robert A. Blake, Jr.[2]*

Criticality of Disbonds in
Laminated Composites

REFERENCE: Chatterjee, S. N., Pipes, R. B., and Blake, R. A., Jr., **"Criticality of
Disbonds in Laminated Composites,"** *Effects of Defects in Composite Materials, ASTM
STP 836,* American Society for Testing and Materials, 1984, pp. 161–174.

ABSTRACT: Usefulness of analytical methodologies and nondestructive tests for assessing
criticality of disbonds in laminated composties is examined. Fracture mechanics approach
and critical energy release rate concepts are employed for the prediction of catastrophic
growth due to quasi-static transverse shear loading. Semi-empirical growth laws are considered for modeling slow growth under cyclic loading.

Brief descriptions of the analytical methods utilized for calculation of energy release
rates are given. Sizes and locations of implanted disbonds in graphite/epoxy laminates are
determined from ultrasonic C-scans. Analytical results and quasi-static destructive test data
for beams subjected to three-point bending are correlated for determining the critical value
of energy release rate. Effects of initial flaw size, location, sharpness of disbond tips, as
well as mixed-mode (I and II) loading conditions are studied. Ultrasonic C-scans are also
used for monitoring growth of disbonds in plate-type structures under cyclic transverse
shear. Least square fits are attempted with the disbond growth rate data.

KEY WORDS: composite materials, fatigue (materials), laminated composites, nondestructive evaluation, defect criticality, delamination fracture, graphite/epoxy materials, fatigue crack growth, fracture mechanics

Nondestructive evaluation (NDE) is an attractive tool for assessing criticality
of defects in aircraft structural components. It also provides a basis for defining
regular inspection intervals for a structural element during its service and for
determining the need for mandatory repair. To achieve the desired objective, it
is necessary to assess criticality of defects by utilizing analytical or semi-empirical
methods or both, after the defects are quantified by the use of nondestructive
measurements. Growth of disbonds in laminated composite structural members
appears to be an important phenomenon, and in the opinion of some investigators

[1] Staff scientist, Materials Sciences Corporation, Spring House, Pa. 19477.
[2] Professor and associate scientist, respectively, Center for Composite Materials, University of
Delaware, Newark, Del. 19711.

the most critical one, which can govern the residual strength or lifetime of such members. Disbonds can initiate near structural discontinuities like free edges or ply drops, may be found as "birth defects," or can be created due to foreign object impact. Growth of such disbonds in Mode I, Mode II, or combined modes under quasi-static and cyclic loading has been modeled by various investigators using critical energy release concepts of linear elastic fracture mechanics and fatigue crack growth laws [1–11].[3] A bibliography of most of the recent studies in this area can be found in Ref 12. Validity of these methodologies, however, has not been fully demonstrated. Further, in all these studies, growth of implanted or inherent disbonds is either one-dimensional or the correlation is attempted via one-dimensional models. This work reports the results of a comprehensive test plan and data correlation studies for testing the validity of the analytical methodologies for prediction of criticality of two types of disbonds: (1) rectangular shaped disbonds in beam type laminates having the same width as that of the beam; and (2) disbonds with elliptic plan forms in laminated plates.

Presence of a disbond does not affect the structural performance of a laminated beam or plate member under in-plane loads except when such loads can cause local buckling due to in-plane compression or shear or both. Under transverse shear, catastrophic or slow growth of disbonds may occur due to quasi-static or cyclic loading, respectively. Details of studies on such growth patterns are reported in the next three sections. Studies on buckling of disbonded laminates under compressive loads are not given here, but are reported in Refs 12 and 13. For brevity, details of the analytical models used in data correlation studies are not described here. Brief descriptions of these models are given next.

Strain energy release rates (or stress intensity factors) at the tips of disbonds located in beam-type members are obtained from a two-dimensional (plane stress) elasticity analysis, where each lamina is considered as an anisotropic material characterized by its effective elastic properties. Details of the stress analysis that reduces the problem to the solution of a set of singular integral equations and numerical solutions of these equations can be found in Ref 11.

Elliptic disbonds in plate type structural members are modeled as two laminated plates bonded everywhere except over the area of the disbond. A laminated plate theory with the effects of surface tractions and shear deformation is used for each plate and the problem is reduced to the solution of a set of integral equations. The model, as described in Ref 13, yields Mode I, Mode II, and Mode III contributions to energy release rates along the disbond periphery. Because of the use of laminated plate theory, the solution is approximate and does not yield the inverse power type singularity, which exists in elasticity solution. The singularity appears in the form of concentrated forces at the disbond tips. The energy release rates, however, should be close to those obtained from more rigorous elasticity solution for disbonds of large size as compared to the thickness

---

[3] The italic numbers in brackets refer to the list of references appended to this paper.

of the plates. No elasticity solution of the problem under consideration is reported
in literature.

Disbonds in Beams—Effects of Location

Tests were performed on three types of laminates of AS-3501 graphite/epoxy
material with seven different defect (two-ply teflon disbond) locations through
the thickness of the samples. All tests were performed in three-point bending at
room temperature. The three types of laminates used in the test are: (1) Type
A—$[0_8/+45_4/-45_4]_{2s}$, (2) Type B—$[0_8/+45_4/-45_4/0_6/+45_4/-45_4/0_2]_s$, and
(3) Type C—$[0_4/\pm45_2/\mp45_2/0_4]_{2s}$.

Test Method

Testing was performed in an Instron load frame at a rate of 1.27 mm (0.05
in.)/min. Prior to testing, the static test machine was balanced and calibrated
and care was taken to assure that the force was introduced perpendicular to, and
in the center of, each of the 25.4 mm by 254 mm by 64-ply graphite/epoxy
beams. Load was applied using a 12.7-mm-diameter circular steel rod and a
12.7-mm-wide copper pad for uniform load introduction and to avoid stress
concentrations at the center of the beam. Geometry of the test specimens is
shown in Fig. 1. All implanted defects were located in the region of compressive
flexural stress. Ultrasonic C-scans were made on all beams prior to testing and
if catastrophic failure did not occur after a significant load drop, the beams were
rescanned and retested. Thickness and width of each beam and length of each
disbond (from C-scans) were measured. These data, as well as measured failure
loads, are tabulated in Ref 12 where details of fabrication are also given.

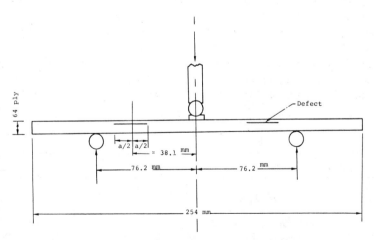

FIG. 1—*Test geometry for three-point bend tests.*

Analysis of Data

All test data were utilized to calculate the critical values of strain energy release rates with the help of the analytical methods developed in Ref *11*. These methods directly yield the stress intensity factors and energy release rates at the tips of two symmetrically located disbonds in a laminated composite beam loaded in three-point bending. The layers are assumed to be in a state of plane stress. Elastic properties of the layers, nondimensionalized by a factor of $E_0 = 6.895$ GPa (10^6 psi), which were used for calculations, are given in Table 1. These values are representative of those measured from experiments [*11*].

Since the $+45$ and -45-deg layers are lumped separately in Laminates A and B, properties listed under Material 2 were used for both of these layers. For consistency, the same properties were also used for Laminate C. Calculations indicate that if the properties of a smeared material (Material 3) are used, the results differ by less than 5%. Laminates A, B, and C are defined in the beginning of this section.

The computer code, DELAM, in its present form [*11*] can solve the problem of disbonds between two unidirectionally reinforced layers of the same orientation. In a few cases, however, the disbonds were between a 0-deg and a $+45$-deg (or a -45-deg) layer. Therefore, calculations were performed for disbonds located between two epoxy matrix layers (Material 4) of equal thickness, which is small compared to the ply thickness. Results were obtained for three values of the matrix layer thickness, and the values of the energy release rates in the limiting case of zero thickness were obtained by extrapolation. The details of this scheme are given in Ref *12*.

It should be noted that for Laminates B and C, the failure loads were also the maximum loads. Some specimens of Laminate A, however, showed some load drops before final failure occurred, details of which can be found in Ref *12*. In some cases, that is, for some of the specimens of the Series A, the first load drop occurred at very low loads. The reasons for such load drops before failure are not very clear. It should be pointed out, however, that the present analysis does not consider the effect of the tearing mode (G_{III}), which can have significant influence, especially in cases where the $+45$ and -45-deg layers are lumped together. Also, shifting of the disbond location across the thickness of the specimen may yield such load drops. These effects cannot be assessed at this time

TABLE 1—*Elastic properties of the layers.*

Material	Layers	C_{11}/E_0	C_{33}/E_0	C_{13}/E_0	C_{55}/E_0
1	0-deg	18.24	1.46	0.409	0.841
2	$+45$ or -45-deg	2.16	1.49	0.294	0.711
3	$(+45/-45)$-smeared	2.88	1.50	0.130	0.687
4	epoxy matrix	0.605	0.605	0.228	0.188

but should be considered in future studies. For the current study, the value of the maximum load was used for calculations. Since the disbond geometry might have changed before the maximum load was attained, the results for Laminate A given in Ref 12 should be interpreted with some caution.

All calculations were performed for a laminate thickness of 8.95 mm and a beam width of 25.4 mm. Experimental values of loads were normalized with due consideration to specimen thickness and width. In the analysis of all test results as well as consideration of the scatter of experimental data that exists in results for three test specimens of the same type, it appears that the critical energy release rate does not vary to a great extent with its location, that is, the orientation of the layers between which the disbond is located. This is expected if one considers the fact that there is always a matrix rich region between two laminae and contends that G^c is a characteristic material constant for the matrix material, which may, of course, depend on its type and processing variables.

To examine the usefulness of the analytical methodology in assessing criticality of disbonds placed at different locations, use was made of the simplest and most commonly employed fracture criterion based on a critical value of total energy release rate to correlate predictions with the experimental data. Failure loads, P_f, for Laminate C predicted by this criterion, that is, $G = G_I + G_{II} = G^c = 1000$ N/m, are plotted in Fig. 2. Similar plots for Laminates A and B are given in Ref 12. These plots show reasonable correlation with experimental data that have been normalized for 2.54-cm-wide beams with 2.54-cm size defects. In

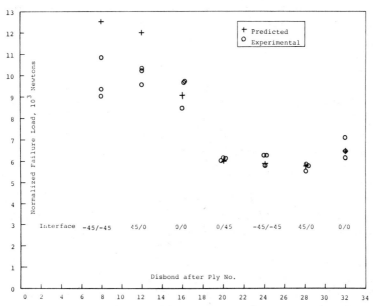

FIG. 2—*Failure loads for disbonds of 2.54-cm-length in 2.54-cm-wide beams—Type C laminates.*

general, the differences between predicted and experimental values are higher when contributions of G_I to the total energy release rates are relatively high, indicating that a quadratic interaction criterion of the form $(G_I/G_I{}^c)^2 + (G_{II}/G_{II}{}^c)^2 = 1$ or even the assumption of linear interaction with different coefficients associated with G_I and G_{II} may yield better correlation with experimental data. No detailed examination of such criteria was attempted, however, since the scatter in test data may not yield a realistic assessment of $G_I{}^c$. Alternative experiments with G_I dominating the response should be designed for this purpose.

It should be pointed out that in cases when $G_I \neq 0$, the stress intensity factor, K_I, at one of the tips is positive, whereas at the other tip, it is negative, indicating existence of compressive stresses near that tip. Therefore, it is quite possible that the two surfaces of the disbond may come in contact near such a tip causing frictional resistance. The effect of this phenomenon cannot be assessed here, but obviously, it may have some influence on the results presented.

Effect of Sharpness of Disbond Tips

Test Method

The test method was the same as that described in the previous section. The 25.4 mm by 254 mm by 64-ply graphite/epoxy beams with defects implanted between the mid-plies were loaded in three-point bending. Maximum load was recorded on a chart recorder. Twelve samples from the test program conducted under a previous study [13], which were 76.2 mm by 254 mm by 64 plies, were used to make the sharp crack specimens. These plates had all been previously fatigued in three-point bending. Cyclic loading is known to yield sharp crack fronts. These beams were all ultrasonically C-scanned to determine the sizes and locations of defects. The 76.2-mm-wide samples were marked and 25.4-mm-wide beams were cut from the center region of each plate. These beams were then rescanned and defect sizes, as well as width and thickness, were measured. These scans are given in Fig. 3 and tabulated results can be found in Ref *12*. Average defect size, radius of the crack front, and crack spacing were determined. These data were used to fabricate the blunt crack specimens. The blunt crack specimens were made from AS-3501-6 prepreg, and the delaminations were implanted between the center plies of the $[0_4/\pm45_2/\mp45_2/0_4]_{2s}$ laminates. The average defect length was made to be 53.98 mm with a crack front radius of 19.05 mm. The defects were made from two-ply teflon.

Smeared $(+45/-45)$ material properties (Material 3 described in the previous section) were used for calculation of $G_{II}E_0\ell/P^2$ $(G_I = 0)$ for midplane disbonds in laminates with a thickness of 8.95 mm and a width of 25.4 mm. Normalized values of failure loads, with due consideration to width and thickness of the specimens, were used for determination of $G_{II}{}^c$. Results are given in Table 2. Results show some scatter and about 10% difference in average $G_{II}{}^c$ for blunt and sharp tips. Considering the scatter, this difference is not significant, and it appears that the nature of the tip does not have any dominant influence on

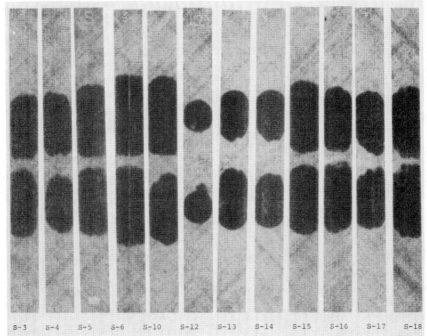

FIG. 3—*C-scans of sharp crack specimens.*

catastrophic disbond growth under static shear. Results from a follow-on study yet to be published [*14*] on blunt disbonds of varying lengths appear to yield further support to this hypothesis.

Elliptic Disbonds in Plates Under Cyclic Transverse Shear

Test Method

Effects of size and location of disbonds placed in $[0_4/\pm45_2/\mp45_2/0_4]_{2s}$ AS-3501-6 graphite/epoxy laminates on flaw growth under cyclic transverse shear were studied in these tests. This was a continuation of the studies on disbonds with two-dimensional planforms reported in Ref *13*. Circular disbonds of 25.4 mm, 31.75 mm, and 38.1 mm diameter in the form of Teflon films were placed in 76.2-mm-wide by 254-mm-long specimens. Two similar disbonds were placed symmetrically in the net section span of 152.4 mm with their center 101.6 mm apart. All fatigue specimens were tested in three-point bending in a manner similar to the beam specimens described in the previous sections. Some of the disbonds were placed in the midplane, that is, between plies 32/33 and others were placed at $Z = \pm0.25H$ (between plies 16/17 or 48/49), H being the laminate thickness. It should be pointed out that $Z = +0.25H$ indicates that the disbonds were in the region of compressive flexural stress and $Z = -0.25H$ means that they were in the region of tensile flexural stress.

TABLE 2—Calculated values of critical energy release rates for blunt and sharp tips.

	Blunt				Sharp		
Sample	Normalized Maximum Load, N	$G_{II}E_0t/P^2$	G_{II}^c, N/m	Sample	Normalized Maximum Load, N	$G_{II}E_0t/P^2$	G_{II}^c, N/m
B-1	4003	61.23	1447	S-3	2681	76.16	807
B-2	3434	62.96	1095	S-4	2964	68.43	887
B-3	3514	64.72	1179	S-5	2833	80.29	951
B-4	3641	62.96	1231	S-6	2783	110.1	1258
B-6	3363	62.96	1050	S-10	2630	70.38	718
B-7	3545	64.72	1200	S-12	5570	26.61	1218
B-9	3316	64.72	1050	S-13	2975	61.23	799
B-10	3315	64.72	1049	S-14	3465	44.94	796
B-11	3060	64.72	894	S-15	2932	78.18	991
B-12	2694	64.72	693	S-16	3119	80.29	1152
B-13	3030	64.72	876	S-17	3028	66.46	899
B-14	2756	66.46	745	S-18	2723	84.85	928
		Average	1042 ± 214			Average	950 ± 175

The samples with 25.4-mm-diameter defects were first tested at an S-level (ratio of maximum load to static ultimate) of 0.5 based on a failure load of 29 400 as determined from static tests reported in Ref *13*. It should be noted that those static failures were due to flexure and not due to transverse shear. Larger disbond dimensions are required for inducing quasi-static disbond growth as observed in an on-going study [*14*]. The specimens with defects located at $Z = +0.25H$ showed little propagation after 12 million cycles. The S-level was, therefore, increased to 0.75 for 31.75-mm and 38.1-mm defects located at $Z = +0.25H$. The S-level was kept at 0.5 for all other samples. The ratio, R, of minimum to maximum load was always kept equal to 0.1. Details of all specimens tested with S-levels, measured thickness and widths are given in Ref *12*.

On an average, the 25.4-mm-diameter defect samples were scanned about 24 times during their life, while the same figure for 31.75-mm- and 38.1-mm-diameter defect samples was about 14. Representative C-scans for one of the samples are given in Fig. 4. Growth of disbonds at three locations on each of the four fronts were measured from each of the scans and the results were tabulated. These data were later used for data correlation studies that are reported in the next section.

Analysis of Data

Strain energy release rates along the periphery of elliptic disbonds of various sizes with semi-axes L_1 and L_2 located at the midsurface of $[0_4/\pm45_2/\mp45_2/0_4]_{2s}$

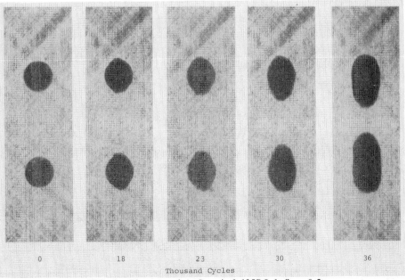

FIG. 4—*Flaw growth in Sample 3-125RO-1, S = 0.5.*

AS-3501-6 graphite/epoxy laminates are reported in Ref *13*. These results were also utilized in the present study. Properties of the unidirectionally reinforced composite (0-deg layer) and stiffness of the $(0_4/\pm45_2/\mp45_2/0_4)_s$ laminate can be found in Refs *12* and *13*. As indicated in Ref *13*, D_{16} and D_{26} were neglected for the calculation and the laminate was considered as orthotropic. Since the disbond is located in the midplane, the energy release rate, G_I, in opening mode is identically equal to zero in this case. In the present study, disbonds were also placed at $Z = \pm0.25H$, H being the laminate thickness. For such location, G_I is nonzero and the variations of G_{II}, G_{III}, and G_I along the disbond periphery are shown in Fig. 5 for some particular geometric configurations. The calculations were performed with laminate properties given in Ref *12*, for self-equilibrating uniform shear stress in x-direction on the disbond surfaces. The two laminates were considered orthotropic, that is, $A_{16} = A_{26} = D_{16} = D_{26} = 0$ for the purpose of calculations. As in the case of disbonds in the midplane, G_{II} first increases with L_1 and then starts decreasing provided L_2 is kept fixed. Similar behavior is also observed for G_I.

Growth of disbonds under cyclic loading observed experimentally were measured and tabulated as described in the previous section. Growth in y-direction (in the direction of the width of the plate specimens) was not significant. Growth rates in x-direction were measured at six points ($y = 0$ and $y = \pm9.53$ mm). The distance L_1 of the two disbond fronts at $y = 0$ from the center of each of the two flaws were plotted against N, the number of cycles for each specimen.

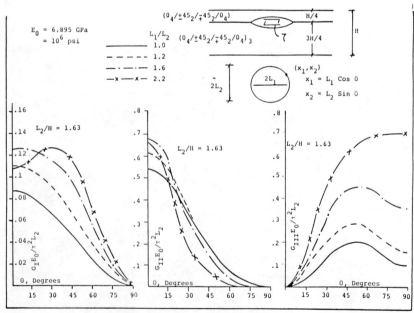

FIG. 5—*Variation of* G_I, G_{II}, G_{III} *with dimensions for flaws at* $Z = \pm0.25H$.

Some representative data are shown in Fig. 6. Growth rates $\Delta L_1/\Delta N$ were then computed at various points along the L_1 versus N plots. For midplane disbond, values of the current total length of the disbond in x-direction were then used as the dimension of the major axis of an approximated elliptic disbond to calculate $\Delta(\sqrt{G_{II}})$, the range of variation of $\sqrt{G_{II}}$. Values of dL_1/dN were then plotted against $\Delta(\sqrt{G_{II}})$ on log-log graph papers.

Results for disbonds in midplane are shown in Fig. 7 along with a semi-empirical curve fit, which is also shown in Fig. 7. In the equation utilized for curve fitting, Δ_0 is the threshold value of $\Delta(\sqrt{G_{II}})$ below which disbond propagation is not expected, and G^c is the critical energy release rate for quasi-static propagation. Several values of Δ_0 and G^c were considered and C_1 and C_2 were obtained from least square fit. Best value of Δ_0 and G^c were then chosen.

Figure 8 shows the plots for disbonds at $Z = \pm0.25H$ along with the semi-empirical growth law obtained for midplane disbonds (Fig. 7) with G_{II} being replaced by $G_I + G_{II}$.

The correlation is not satisfactory especially at the upper and lower ends of the spectrum. Although the data points at the upper end are mostly for $S = 0.75$, effect of mean stress does not enter the picture, since R ratio (minimum stress/maximum stress) is the same. Possible reasons for the unsatisfactory correlation are: (1) comparison with plane strain results [12] shows that the computer code, FLAW, used to calculate the energy release rates, G_I and G_{II}, overestimates G_I and underestimates G_{II}, which might explain the apparent horizontal shift of the experimental data points (Fig. 8) to the rights; and (2) it is possible that the parameters appearing in the growth laws are different for Mode I and Mode II, since the growth rates under Mode I may be different from those under Mode II.

Before closing this discussion, it should be pointed out that growth of a disbond under pure Mode II loading should be the same at $x = \pm L_1$ as predicted by the

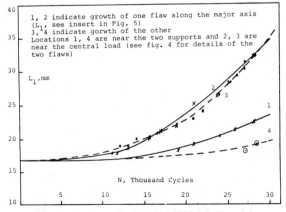

FIG. 6—L_1 versus N, Sample 3-125RO-2, S = 0.5.

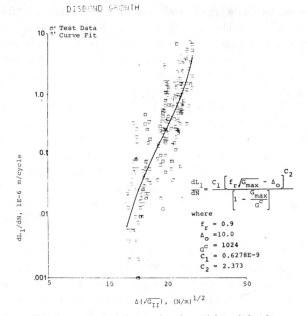

FIG. 7—*Curve fit to fatigue data for midplane disbonds.*

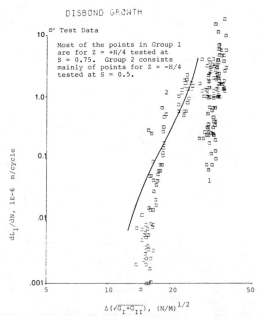

FIG. 8—*Fatigue data for disbonds at* Z = ±H/4.

empirical growth law and the stress analysis model used here. The results for beams with midplane disbonds, however, show some differences between the growths at the four fronts, possibly because of the presence of inhomogeneities or the approximations or both used in the analysis. This is one of the reasons for the scatter noticed in data correlation. Presence of mixed mode conditions for disbonds at $Z = \pm0.25H$ may introduce larger differences because of the existence of tensile and compressive normal stresses at the two fronts. Fatigue data from this study and those from Ref *13* are being compared in an ongoing investigation [*14*].

Discussion and Conclusions

From the results reported herein, the following conclusions and comments can be made.

1. A linear elastic fracture mechanics approach and methods of stress analysis based on two-dimensional elasticity theory can be used to assess criticality of disbonds in laminated beam-type structures under transverse shear, irrespective of disbond size and location. Validity of the proposed methodology has been demonstrated for cases where Mode II energy release rate is the dominant factor influencing failure. Further work appears necessary to assess the effects of Mode III release rate as well as the interaction of all three modes; that is, Modes I, II, and III, in causing failure.

2. Sharpness of disbond tips introduced by fatigue loading does not appear to have a significant influence on $G_{II}{}^c$, as seen from comparison with test data from disbonds with blunt tips, that is, implanted teflon defects. However, because of the scatter in experimental results, more tests are needed before a final conclusion is drawn.

3. Methods of stress analysis based on laminated plate theory yield reasonably accurate values of energy release rates along the periphery of disbonds in laminated plate-type structures under transverse shear. Usefulness of semi-empirical crack growth laws in predicting the growth of such flaws, for cases where Mode II energy release rate is dominant, has been clearly illustrated. Further work for evaluating effects of varying mean stress as well as influence of Mode I and Mode II energy release rates on flaw growth are needed to put the NDE methodology for assessing criticality of disbonds under fatigue loading on a sound basis. Detailed examination of the fracture surface may indicate the relative influences of Mode I and Mode II, since dewetting of fibers along the surface is usually expected in Mode I dominated fracture for high fiber volume fractions. Such examinations were not attempted in this study.

Acknowledgment

The results reported here were obtained from research supported by Naval Air Development Center Contracts N62269-78-C-0111, N62269-79-C-0209, N62269-

80-C-0271, and N62269-81-C-0271. The authors express their sincere appreciation to Dr. W. R. Scott and Dr. B. W. Rosen for numerous suggestions and help during various stages of this work.

References

[1] Bascom, W. D., Bitner, J. L., Moulton, R. J., and Siebert, A. R., *Composites Journal*, 1980, pp. 9–18.
[2] de Charentenay, F. X. and Benzeggagh, M., "Fracture Mechanics of Mode I Delamination in Composite Materials," *Proceedings*, International Conference on Composite Materials 3, Paris, France, Aug. 1980.
[3] Devitt, D. F., Schapery, R. A., and Bradley, W. L., *Journal of Composite Materials*, Vol. 14, Oct. 1980, pp. 270–285.
[4] Wilkins, D. J., Eisenmann, J. R., Camin, R. A., Margolis, B. S., and Benson, R. A., "Characterizing Delamination Growth in Graphite-Epoxy," presented at the ASTM Symposium on Composites Damage Tolerance, Miami, Nov. 1980.
[5] Ramkumar, R. L., Kulkarni, S. V., Pipes, R. B., and Chatterjee, S. N. in *Fracture Mechanics, ASTM STP 677*, American Society for Testing and Materials, 1979, p. 668.
[6] Crossman, F. W., Warren, W. J., Wang, A. S. D., and Law, G. E., "Initiation and Growth of Transverse Cracks and Edge Delamination in Composite Laminates, Part 2, Experimental Correlation," Drexel University, Phila., Pa., 1979.
[7] Rybicki, E. F., Schmueser, D. W., and Fox, J., *Journal of Composite Materials*, Vol. 11, 1977, p. 470.
[8] Wang, S. S., *ASME Journal of Applied Mechanics*, Vol. 47, 1980, pp. 64–70.
[9] Wang, S. S., *Composite Materials: Testing and Design (Fifth Conference), ASTM STP 674*, American Society for Testing and Materials, 1979, pp. 642–663.
[10] Ratwani, M. M. and Kan, H. P., "Compression Fatigue Analysis of Fiber Composites," NADC-78049-60, Naval Air Development Center, Warminster, Pa., Sept. 1979.
[11] Chatterjee, S. N. in *Modern Developments in Composite Materials and Structures*, J. R. Vinson, Ed., American Society of Mechanical Engineers, New York, 1979, p. 1; also see Chatterjee, S. N., Pipes, R. B., and Hashin, Z., "Definition and Modelling of Critical Flaws in Graphite Fiber Reinforced Resin Matrix Composite Materials," NADC-77278-30, Naval Air Development Center, Warminster, Pa., Aug. 1979.
[12] Chatterjee, S. N. and Pipes, R. B., "Composite Defect Significance," Warminster, Pa., NADC-80048-60, Naval Air Development Center, Aug. 1981.
[13] Chatterjee, S. N. and Pipes, R. B., "Study of Graphite/Epoxy Composite for Material Flaw Criticality," NADC-78241-60, Naval Air Development Center, Warminster, Pa., Nov. 1980.
[14] Chatterjee, S. N., Pindera, M. J., Pipes, R. B., and Dick, B., "Composite Defect Significance," NADC-81034-60, Naval Air Development Center, Warminster, Pa., Nov. 1982.

John D. Whitcomb[1]

Strain-Energy Release Rate Analysis of Cyclic Delamination Growth in Compressively Loaded Laminates

REFERENCE: Whitcomb, J. D., "**Strain-Energy Release Rate Analysis of Cyclic Delamination Growth in Compressively Loaded Laminates,**" *Effects of Defects in Composite Materials, ASTM STP 836,* American Society for Testing and Materials, 1984, pp. 175–193.

ABSTRACT: Delamination growth in compressively loaded composite laminates was studied analytically and experimentally. The configuration used in the study was a laminate with an across-the-width delamination. An approximate superposition stress analysis was developed to quantify the effects of various geometric, material, and load parameters on Mode I and Mode II strain-energy release rates G_I and G_{II}, respectively. Calculated values of G_I and G_{II} were then compared with measured cyclic delamination growth rates to determine the relative importance of G_I and G_{II}. High growth rates were observed only when G_I was large. However, slow growth was observed even when G_I was negligibly small. This growth apparently was due to a large value of G_{II}.

KEY WORDS: composite materials, fatigue (materials), fracture mechanics, laminates, delamination, local buckling, nonlinear stress analysis, compression

Nomenclature

a — Half-length of delamination before loading

$\bar{a}$ — Half-length of delamination after loading

Δa — Virtual crack closure distance used in strain-energy release rate calculations

b — Specimen width

$C, C1, C2, C3,$ $n, n1, n2, Z$ — Arbitrary constants

D — Bending stiffness of the buckled region given by

$$D = \frac{b}{3} \sum_{k=1}^{\rho} E^k \left[\left(\alpha_k - \frac{t}{2} \right)^3 - \left(\alpha_{k-1} - \frac{t}{2} \right)^3 \right]$$

[1] Research engineer, NASA Langley Research Center, Hampton, Va. 23665.

where ρ = number of plies

d_x, d_y	Unit load solutions for displacements near crack tip
E^k	Young's modulus for ply k
E_{11}, E_{22}, E_{33}	Young's moduli of unidirectional ply. The subscripts 1, 2, and 3 refer to the longitudinal, transverse, and thickness directions, respectively.
F_x, F_y	Unit load solutions for forces at crack tip
G_I	Mode I strain-energy release rate
G_{II}	Mode II strain-energy release rate
$\hat{G}_I$	Maximum possible value of G_I for current delamination length
G_{12}, G_{13}, G_{23}	Shear moduli of unidirectional ply
M	Moment
N	Number of applied load cycles
P_A, P_B, P_C, P_D	Axial loads in Regions A, B, C, and D, respectively
P_T	Remote applied compressive load
S_A, S_B, S_C, S_D	Axial stiffness of Regions A, B, C, and D given by

$$S = b \sum_{k=1}^{\rho} E^k(\alpha_k - \alpha_{k-1})$$

where ρ = number of plies

t	Thickness of buckled region
x, y	Rectangular Cartesian coordinates
α_{k-1}	Distance from top surface of laminate to ply k; top ply is Ply 1
δ	Lateral deflection at $x = -a$ due to applied load
$\hat{\delta}$	Value of δ corresponding to $\hat{G}_I$
δ_0	Initial lateral deflection at $x = -a$
ν_{12}, ν_{13}, ν_{23}	Poisson's ratios for unidirectional ply

In composite structures subjected to compression loads, delaminations can cause localized buckling (Fig. 1). High interlaminar stresses at the edges of the

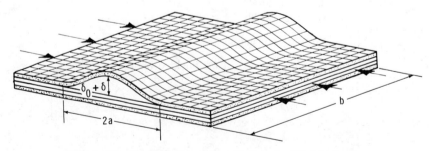

FIG. 1—*Local buckling of laminate with through-width delamination.*

buckled region often lead to cyclic delamination growth (herein referred to as instability-related delamination growth).

The objective of this paper is to investigate the mechanism of instability-related delamination growth. Figure 1 shows the configuration used in the study—a laminate with a "through-width" delamination. This configuration was selected because it is perhaps the simplest configuration that exhibits instability-related delamination growth. Goals of the investigation were: (1) to develop and use an approximate superposition stress analysis to explain how various geometric, material, and load parameters affect interlaminar stresses, (2) to determine the delamination growth behaviors predicted by several different criteria based on strain-energy release rates, and (3) to compare analytical calculations with experimental observations to determine the applicability of each growth criterion.

Because of the stress singularity at the end of the delamination (crack tip), calculated stresses there have little meaning. Strain-energy release rates are finite parameters that characterize the intensity of the stresses near the crack tip. Consequently, in the following discussion, strain-energy release rates will be used to characterize the severity of the interlaminar stresses.

Analysis

The configuration shown in Fig. 1 was idealized as a two-dimensional plane-strain problem. Linear and nonlinear finite-element analyses and an approximate superposition analysis were used to calculate strain-energy release rates for the two-dimensional idealization. The nonlinear analysis was used to provide reference solutions for evaluation of the approximate superposition analysis. The linear analysis was used to calculate several constants used in the approximate superposition analysis. The nonlinear analysis is described in Ref 1,[2] and the linear analysis is simply a linear version of this analysis.

The approximate superposition analysis, the procedure for calculating strain-energy release rates, the finite-element models, and material properties are discussed in the following sections.

Approximate Superposition Analysis

Superposition techniques have been widely used in linear stress analysis to represent a complicated problem as a combination of several simpler problems. Application of the principle of superposition to nonlinear problems first requires a transformation that results in a linear system.

The key to the transformation is replacement of the source of nonlinearity with equivalent loads (Figs. 2a and 2b). Because of symmetry, only half of the configuration is considered. The buckled region (which responds nonlinearly due to significant rotations) is replaced by the loads, P_D and M, the axial load and moment, respectively, in the column where it is cut (Fig. 2b). The new

---

[2] The italic numbers in brackets refer to the list of references appended to this paper.

configuration is linear, with three nonlinearly related applied loads, P_T, P_D, and M. By superposition, the number of loads can be reduced to two, as illustrated in Figs. 2c through 2e.

The load system in Fig. 2c is divided into the two load systems shown in Figs. 2d and 2e. Because P_C is calculated using rule of mixtures, the load system in Fig. 2e causes a uniform axial strain state and no interlaminar stresses. Consequently, in terms of interlaminar stresses, only the load system in Fig. 2d (that is, $(P_C - P_D)$ and M) need be considered. Accordingly, in the current study involving strain-energy release rates, Fig. 2d is the linearized equivalent of the nonlinear problem in Fig. 2a.

The appendix describes a strength of materials analysis for calculating $(P_C - P_D)$ and M. The key equations from the appendix are

$$P_T = \frac{\pi^2 S_A}{a^2} \left[\frac{(\delta^2 + 2\delta\delta_0)}{16} + \frac{D(S_A + S_D)}{S_A S_D} \frac{\delta}{\delta + \delta_0} \right] \tag{1}$$

$$P_C - P_D = \frac{S_D}{S_A + S_D} P_T - \frac{\pi^2 D}{a^2} \frac{\delta}{\delta + \delta_0} \tag{2}$$

$$M = \frac{\pi^2 D}{2a^2} \delta \tag{3}$$

To use the loads $(P_C - P_D)$ and M in a two-dimensional analysis requires that they be expressed as an equivalent distribution of tractions. To calculate this distribution, the axial strains were assumed to vary linearly through the thickness where the tractions are applied (that is, at the cut). Intuitively, this seems to be reasonable if Region D (Fig. 2) is not cut too close to the crack tip. The validity of the assumed linear variation will be checked later in this paper.

Linear finite-element analysis was used to calculate the response of the lin-

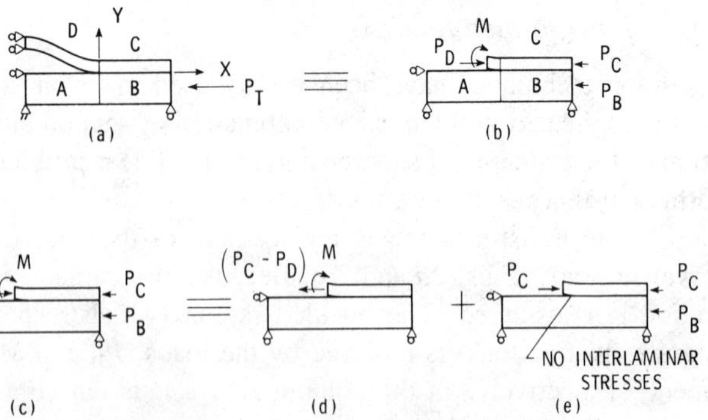

FIG. 2—Nonlinear configuration (a) transformed into linear configuration (d) with two nonlinearly related loads, $(P_C - P_D)$ and M.

earized configuration in Fig. 2d to unit values of $(P_C - P_D)$ and M. Because the configuration is linear, the solution for any arbitrary combination of $(P_C - P_D)$ and M is simply a linear combination of the unit load responses. If Region B (Fig. 2) is much thicker than Region C, the unit load solutions are very insensitive to delamination length. In the current study, the ratio of thicknesses was 61 to 3. Hence, the unit load solutions for $2a = 25$ mm were used for analyzing all delamination lengths. Also, initial waviness of the buckled region does not enter into the finite-element analysis. Delamination length and initial waviness were both accounted for in the strength of materials analysis in calculating $(P_C - P_D)$ and M, Eqs 2 and 3, respectively. This procedure will be discussed further in the next section.

Strain-Energy Release Rate

The virtual crack closure method (Ref 2) was used to calculate Mode I and Mode II strain-energy release rates, G_I and G_{II}, respectively. The forces transmitted through the node at the crack tip and the relative displacements of the two nodes on the crack boundary closest to the crack tip were used in the calculation. Equation 4 shows how this technique is used for the superposition stress analysis.

$$G_I = \frac{1}{2\Delta ab} [(P_C - P_D)F_y^1 + MF_y^2] [(P_C - P_D)d_y^1 + Md_y^2]$$

$$G_{II} = \frac{1}{2\Delta ab} [(P_C - P_D)F_x^1 + MF_x^2] [(P_C - P_D)d_x^1 + Md_x^2]$$

(4)

In these equations, F_x, F_y, d_x, and d_y are the unit load values of the nodal forces and the corresponding relative nodal displacements in the x and y directions. (The coordinate system is defined in Fig. 2.) Superscripts 1 and 2 on the unit load parameters identify parameters associated with $(P_C - P_D)$ and M, respectively.

If the distance is small between the crack tip and the nodes used to calculate relative displacements, then $F_y^1/d_y^2 = F_y^2/d_y^2$ and $F_x^1/d_x^1 = F_x^2/d_x^2$. Using these relationships in Eq 4 results in

$$G_I = \frac{1}{2\Delta ab} \frac{d_y^1}{F_y^1} [(P_C - P_D)F_y^1 + MF_y^2]^2$$

$$G_{II} = \frac{1}{2\Delta ab} \frac{d_x^1}{F_x^1} [(P_C - P_D)F_x^1 + MF_x^2]^2$$

(5)

In the Results and Discussion section of this paper, it will be shown that for high loads or long delamination lengths, G_I is zero, that is, the crack tip closes in the normal direction. To prevent the crack faces from overlapping (analytically) requires the addition of multipoint constraints on the crack face nodes. Conceptually, the crack face nodes are connected in the direction normal to the crack face by infinitesimal springs. These springs have infinite stiffness in compression and zero stiffness in tension. To determine whether to select zero or infinite stiffness requires solution of a nonlinear contact problem. To include the contact problem directly in the superposition analysis would severely complicate the otherwise simple equations. Therefore, use of a non-contact analysis to approximate G_{II} was investigated.

A laminate with $2a = 76.2$ mm was analyzed using two different approaches. First, contact forces were ignored (that is, overlap of crack faces was allowed), and G_I and G_{II} were calculated using Eq 5. In the second approach, overlap of the crack faces was prevented, which is more realistic, and G_{II} was calculated using Eq 5. (Note that Eq 5 yields $G_I = 0$ when overlap is prevented.) Applied loads (P_T) ranged from 14.8 kN, which corresponds approximately to initial crack tip closure, to 55.2 kN.

When crack face overlap was prevented, a larger value of G_{II} was calculated than when overlap was allowed. The difference in the G_{II} values increased with load. But in all cases, the difference was approximately equal to G_I calculated using the approach that allowed crack face overlap. For example, for $P_T = 55.2$ kN, the contact analysis yielded $G_{II} = 413$ J/m². When crack face overlap was allowed, G_I and G_{II} were 35 and 384 J/m², respectively. The sum of these values is within approximately 1.5% of the more realistic solution, that is, $G_{II} = 413$ J/m². Apparently, the crack-face contact forces do not significantly alter the total strain-energy release rate. Hence, when there is crack tip closure, the total strain-energy release rate from the non-contact analysis can be used to approximate G_{II} (which is then the total strain-energy release rate, since G_I is identically zero).

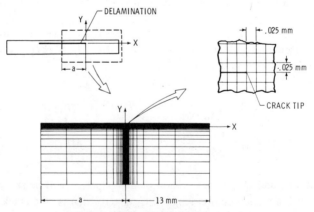

FIG. 3—*Typical finite-element mesh.*

Finite-Element Model

A typical finite-element mesh for the nonlinear analysis is shown in Fig. 3. Because of symmetry, only half of the laminate was modeled. The mesh contains 813 nodes and 740 four-node isoparametric elements. Reduced integration was used to improve the performance of the elements in modeling bending deformations. Because the rotations are small except in part of the buckled region, the nonlinear strain-displacement relationships were used only for the region $y > 0$, $-a \leq x \leq -0.56$ mm. As shown in Fig. 2b, the linearized configuration is the same as the nonlinear configuration except that most of the buckled region is removed. Accordingly, the mesh used in the linear analysis was derived from that in Fig. 3 by removing elements in the deleted part of the buckled region.

Materials Properties

The material studied was NARMCO T300/5208[3] graphite/epoxy. The unidirectional-ply properties were assumed to be

$$E_{11} = 140 \text{ GPa}$$
$$E_{22} = E_{33} = 14 \text{ GPa}$$
$$\nu_{12} = \nu_{13} = \nu_{23} = 0.21$$
$$G_{12} = G_{13} = G_{23} = 5.9 \text{ GPa}$$

These properties were selected because of their wide usage in other analytical studies. Plane strain (that is, $\epsilon_z = 0$) and $\epsilon_{xz} = 0$ were imposed to calculate the two-dimensional properties. In regions where coarse finite elements spanned several plies, laminate theory was used to obtain average properties.

Experimental Procedure

The specimens used for this study were fabricated and tested by Northrop Corporation. (Details are in Ref 3.) A cursory description of the experimental procedure is given herein.

The specimen consisted of 64 plies of T300/5208. The fiber orientation and stacking sequence were $[0_4/(0/45/90/-45)_7]_s$. The laminate width, b, was 25.4 mm. To simulate a delamination, kapton film was used to prevent bonding over a 19-mm length between the third and fourth plies. The ply thickness was assumed to be 0.14 mm. Six specimens were tested in fatigue under compressive constant-amplitude loads. Minimum compressive load was 10% of the maximum compressive load. The load frequency was 10 Hz. Delamination lengths were measured with a microscope.

---

[3] Use of trade names or manufacturers does not constitute an official endorsement, either expressed or implied, by the National Aeronautics and Space Administration.

Results and Discussion

First, the accuracy of the approximate superposition analysis will be evaluated. Then the effect of various parameters on G_I and G_{II} will be considered. Finally, the experimental observations will be compared with the analytical results.

Evaluation of Approximate Superposition Analysis

The approximate superposition analysis was evaluated by comparison with results from a geometrically nonlinear finite-element analysis. Recall that a major assumption in the approximate analysis was that the strains vary linearly through the thickness where $(P_C - P_D)$ and M are applied. Figure 4 shows the axial strain variation through the thickness at $x = -0.2$ and -0.7 mm obtained using the nonlinear finite-element analysis. Along the line $x = -0.7$ mm, the strains vary almost linearly for the three applied loads. However, closer to the crack tip along $x = -0.2$ mm, the variation is more nonlinear, especially near $y = 0$. In the following, $(P_C - P_D)$ and M were applied at $x = -0.76$ mm. The virtual crack closure distance, Δa, was 0.0254 mm.

The unit load solutions F_x, F_y, d_x, and d_y are

$$F_x^1 = 9.36 \times 10^{-2}$$
$$F_x^2 = 531 \text{ m}^{-1}$$
$$F_y^1 = 0.0261$$
$$F_y^2 = -252 \text{ m}^{-1}$$
$$d_x^1 = 1.40 \times 10^{-10} \text{ m-N}^{-1}$$
$$d_y^1 = 2.97 \times 10^{-10} \text{ N}^{-1}$$

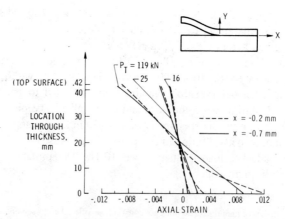

FIG. 4—Variation of axial strain through thickness of buckled region (2a = 25 mm), nonlinear analysis.

These values were used in Eq 5 for any combination of $(P_C - P_D)$ and M to obtain G_I and G_{II}.

Differences between G_I and G_{II} from the approximate superposition analysis and the geometrically nonlinear finite-element analysis can be traced mainly to two sources: (1) nonlinear variation of the strains through the thickness of the buckled region and (2) inaccuracy in determining $(P_C - P_D)$ and M.

By using $(P_C - P_D)$ and M from the geometrically nonlinear finite-element analysis, the effect of nonlinear variation of the strains can be examined. Figure 5 shows that this effect is small.

Figure 6 shows that if the strength of materials analysis is used to calculate $(P_C - P_D)$ and M, the difference is much larger. Hence, most of the difference between the two analyses is due to inaccuracy in determining $(P_C - P_D)$ and M. But the general trends for the G_I variation with delamination length and load are predicted very well. In Fig. 6, the curves for the two analyses seem to differ (approximately) by a constant scale factor.

A direct test of the approximate analysis for predicting trends is to use it to coalesce the curves in Fig. 6 into a single curve. Equations 2 and 3 show that $(P_C - P_D)$ and M can be expressed as functions of δ. Hence, from Eq 5, G_I and G_{II} are functions of δ. Equations 1, 2, 3, and 5 show that for constant δ, P_T varies as a^{-2}, and G_I and G_{II} vary as a^{-4}. Hence, plotting $a^4 G_I$ versus $a^2 P_T$ should coalesce the curves for various delamination lengths. Figure 7 shows that the data for five delamination lengths (including those in Fig. 6) do coalesce into a narrow band around a single curve. Since the peak values of G_I for various lengths differ by more than two orders of magnitude, the closeness of the fit

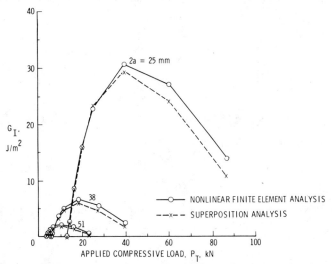

FIG. 5—*Evaluation of approximate superposition analysis* (($P_C - P_D$) *and* M *from nonlinear finite-element analysis*).

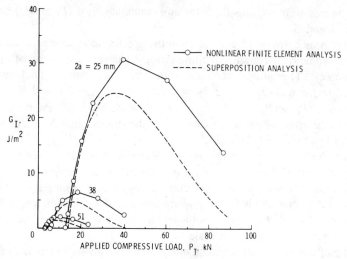

FIG. 6—*Evaluation of approximate superposition analysis* (($P_C - P_D$) *and* M *from strength of materials analysis*).

suggests the approximate analysis is accurate for predicting trends. Therefore, all results that follow are obtained with the approximate superposition analysis. Figure 7 also shows that if nonlinear finite-element results are available for one delamination length, the values for other lengths can be estimated immediately.

An advantage of the superposition analysis is that it allows a problem to be dissected. In particular, one can determine the relative importance of the loads ($P_C - P_D$) and M on G_I and G_{II}. Figures 8 and 9 show G_I and G_{II} calculated by

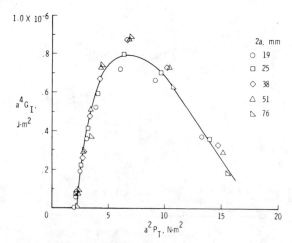

FIG. 7—*Compaction of results from geometrically nonlinear analysis.* (*Curve is visual fit of data.*)

using M alone and by using $(P_C - P_D)$ and M in combination. Although intuition might suggest that only the peeling action caused by the moment, M, has a significant effect on G_I, Fig. 8 shows that $(P_C - P_D)$ contributions cannot be ignored. Figure 9 shows that both $(P_C - P_D)$ and M are also important when calculating G_{II}.

Parametric Study

The effects of several parameters on G_I and G_{II} were examined using the approximate superposition analysis. The parameters were initial waviness, delamination length, applied load, and the ratio of axial to bending stiffness for the buckled region.

Initial imperfections in the form of simple sinusoidal waviness were assumed (Eq 6)

$$v(x)\big|_{\text{initial}} = \frac{\delta_0}{2}\left(1 - \cos\frac{\pi x}{a}\right) \tag{6}$$

where $v(x) = $ distortion in the y direction. When a column is initially wavy, bifurcation buckling does not occur. As soon as load is applied, the column begins to deflect laterally, which causes interlaminar stresses. Hence, G_I and G_{II} are nonzero as soon as load is applied. If $\delta_0 = 0$, G_I and G_{II} are zero until buckling occurs. However, Fig. 10 shows that the peak value of G_I is significantly reduced, even for very small imperfections. In contrast, Fig. 11 shows that G_{II} is hardly affected by initial waviness.

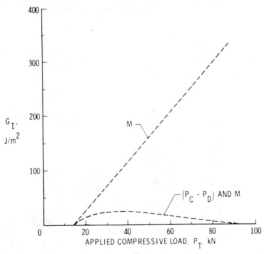

FIG. 8—*Effect of* $(P_C - P_D)$ *and* M *on* G_I *(2a = 25 mm).*

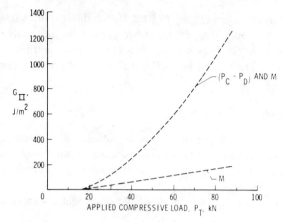

FIG. 9—*Effect of* (P_C − P_D) *and* M *on* G_{II} (2a = 25 *mm*).

Figures 6, 12, and 13 show the effect of delamination length on G_I and G_{II}. The shorter delaminations have the larger values of peak G_I (that is, $\hat{G}_I$) (Fig. 6). However, for the longer delaminations, G_I becomes nonzero at lower loads. Figure 12 shows that after only a little delamination growth, G_I reaches a peak and decreases rapidly with further growth. At $2a$ = 40 to 50 mm, the crack tip closes in the normal direction and G_I is identically zero. Further delamination growth causes compressive normal stresses to develop at the crack tip. In contrast, G_{II} initially increases then decreases only slightly to a constant value with increased delamination length (Fig. 13). Note that G_{II} is typically much larger than G_I.

Figures 6 and 14 illustrate the effects of applied load on G_I and G_{II}, respectively. The Mode I strain-energy release rate, G_I, first increases to a peak value

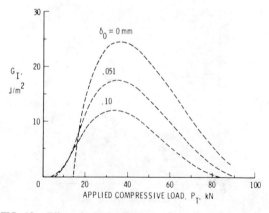

FIG. 10—*Effect of initial imperfection* δ_0 *on* G_I (2a = 25 *mm*).

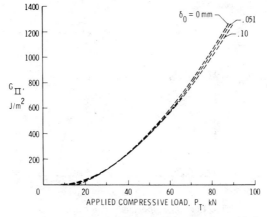

FIG. 11—*Effect of initial imperfection δ_0 on G_{II} (2a = 25 mm).*

$(\hat{G}_I)$, then decreases with increasing load (Fig. 6). In contrast, G_{II} monotonically increases with increasing load. As a result, G_I and G_{II} do not usually reach peak values at the same time during a fatigue load cycle. Furthermore, the load at which G_I is maximum decreases with increasing delamination length.

Since $\hat{G}_I$ is the maximum possible value of G_I for a given delamination length, it is of interest how $\hat{G}_I$ varies with delamination length. The first step in determining this variation is to determine the lateral deflection, $\hat{\delta}$, corrresponding to $\hat{G}_I$. The lateral deflection, $\hat{\delta}$, is obtained by solving

$$\frac{\partial G_I}{\partial \delta} = 0 \tag{7}$$

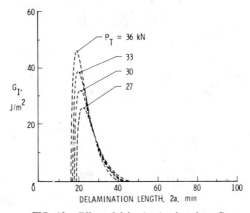

FIG. 12—*Effect of delamination length on G_I.*

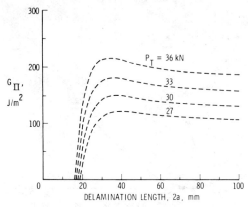

FIG. 13—*Effect of delamination length on G_{II}*.

Equations 2, 3, 5, and 7 are combined to obtain the governing equation (Eq 8)

$$d_y^1 \left[\frac{S_A S_D}{S_A + S_D} \frac{\hat{\delta}(1 + \delta_0)}{8} + \frac{D\delta_0}{(\hat{\delta} + \delta_0)^2} \frac{1 - S_A}{S_A} \right] + \frac{d_y^2 D}{2} = 0 \qquad (8)$$

Equation 8 is solved iteratively for $\hat{\delta}$. Once $\hat{\delta}$ is determined, $\hat{G}_I$ can be calculated from Eqs 1, 2, 3, and 5. Note that $\hat{\delta}$ is independent of delamination length. Earlier it was shown that for constant δ, G_I and G_{II} vary as a^{-4}, and P_T varies as a^{-2}. Hence, $\hat{G}_I$ varies as a^{-4} and, the corresponding applied load varies as

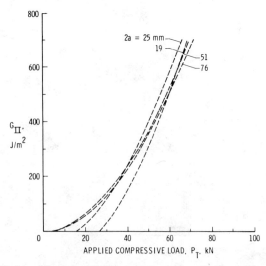

FIG. 14—*Effect of applied compressive load on G_{II} (2a = 25 mm)*.

a^{-2}. The corresponding value of G_{II} also varies as a^{-4}. These observations will be of special interest later when examining the fatigue data.

The last parameter to be examined is the ratio of flexural-to-axial stiffness (that is, D/S_{D}) of the delaminated region. The buckling load for the region is linearly related to D. Prior to buckling, the load in the delaminated region is linearly related to S_{D}. Hence, for thick specimens with a thin delaminated region, the applied load that causes buckling depends on the ratio, D/S_{D}. Delaminated regions are less prone to buckle if they possess low axial stiffness and high flexural stiffness. But for homogeneous materials or unidirectional orthotropic laminates, this ratio is simply

$$\frac{D}{S_{\mathrm{D}}} = \frac{t^2}{12} \tag{9}$$

Hence, the applied load that causes buckling is independent of the material properties of the buckled region. However, for multi-directional laminates the ratio, D/S_{D}, depends on both the lamina properties and the stacking sequence. For example, the value of S_{D} for a $[0_3]$ laminate is approximately 1.4 times that for a $[0/90/0]$ laminate, but the value of D is essentially the same for both laminates. Consequently, for a thick laminate with a delamination, buckling occurs at a lower applied load if the delaminated region is $[0_3]$ rather than $[0/90/0]$.

Comparison of Analysis and Fatigue Data

The roles of G_{I} and G_{II} in delamination growth were investigated by comparing calculated values of G_{I} and G_{II} with measured growth rates. Fatigue data for six specimens from Ref 3 were used for comparison with analytical results. Three of the specimens were tested at $(P_T)_{\mathrm{max}} = 33$ kN and another three at $(P_T)_{\mathrm{max}} = 30$ kN. The results are presented in Fig. 15. The da/dN decreased rapidly with delamination growth. Both curves are approximately linear with a slope of -4, hence

$$\frac{da}{dN} \approx Ca^{-4} \tag{10}$$

Figure 16 shows da/dN versus the maximum values of G_{I} and G_{II} during fatigue cycling at a maximum compressive load of 33 kN. Note that the growth rate is largest when G_{I} is relatively large. But slow growth is observed even when G_{I} is very small. The Mode II strain-energy release rate, G_{II}, changes little after initial delamination growth. Since G_{II} remains large, delamination growth likely is driven by G_{II}.

Two delamination growth criteria were examined to determine whether they could predict the observed growth rates. The first growth criterion examined is

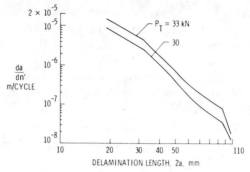

FIG. 15—*Effect of delamination length and load on growth rate.*

given by

$$\frac{da}{dN} = ZG_I^n \tag{11}$$

where Z and n are constants. Equation 11 is evaluated at the point in the load cycle when G_I is maximum. The load range and delamination lengths are such that for almost the entire test, the maximum G_I during each load cycle is $\hat{G}_I$, which was obtained by solving Eq 7. Recognizing that $\hat{G}_I$ decreases as a^{-4}, as shown earlier, Eqs 10 and 11 can be solved for n; the result is $n = 1$. This is in strong contrast to published values of $n \approx 15$ to 20 for double cantilever beam

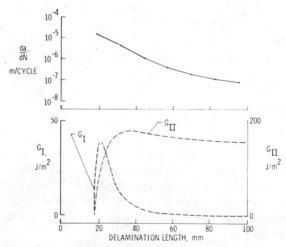

FIG. 16—*Comparison of G_I, G_{II}, and delamination growth rate for maximum fatigue load of 33 kN.*

fatigue tests (Ref *4*). Apparently, Eq 11 is not a valid growth criterion for the specimens considered.

Next a growth criterion was considered that includes both G_I and G_{II}. If we assume there is no synergistic interaction of G_I and G_{II} (that is, the effects are separable), then

$$\frac{da}{dN} = f_1(G_I) + f_2(G_{II}) \tag{12}$$

where f_1 and f_2 are functions of G_I and G_{II}, respectively.

From the double-cantilever beam data just discussed, we know that f_1 is extremely sensitive to G_I (that is, $da/dN \propto G_I^{15}$). Since G_I decreases rapidly with increasing a, f_1 must also decrease extremely fast as a increases. In fact, f_1 would not contribute noticeably to da/dN after the initial growth. Hence, delamination growth appears to be driven by G_{II} alone. Accordingly, it was assumed that the growth criterion should be evaluated when G_{II} is maximum, that is, at peak load. However, earlier it was shown that for long delaminations the crack tip closes and produces compressive σ_y stresses when the cyclic load is maximum. The compressive stress probably reduces the effect of G_{II} on delamination growth, but it was not clear how to account for this stress. Two approaches were tried: (1) ignore the compressive normal stress and set $f_1 = 0$ when the tip closes, or (2) let f_1 take on negative values after the crack tip closes.

If we choose to set $f_1 = 0$ when the crack tip closes, then

$$\frac{da}{dN} = f_2(G_{II}) \tag{13}$$

for virtually the entire test. Figure 14 showed that G_{II} first increased then decreased slightly as the delamination extends. In the experiments, two load levels were used: $(P_T)_{max} = 33$ kN and 30 kN. Figure 14 shows that for $2a > 25$ mm, the minimum value of G_{II} for $(P_T)_{max} = 33$ kN is greater than the maximum value of G_{II} for $(P_T)_{max} = 30$ kN. Hence, Eq 13 would predict that for $2a > 25$ mm, the minimum da/dN for the higher load should exceed the maximum da/dN at the lower load. Figure 16 shows this is not the case. Hence, Eq 13 is not valid.

If we select a function, f_1, that becomes negative when the crack tip closes, then we (analytically) allow compressive normal stresses at the crack tip to retard delamination growth due to G_{II}. Since the compressive crack tip stresses increase as the delamination grows, such a function, f_1, would predict a decrease in growth rate with increased delamination length. Although this prediction agrees with the data trend in Fig. 15, more tests are needed to verify or disprove this interpretation.

Despite the complexity of the growth behavior, two trends were clear: (1)

high growth rates were observed only when G_I was large, and (2) slow growth was observed even when G_I was negligibly small; apparently, G_{II} alone can drive delamination growth.

Conclusions

Analysis and experiments were used to study instability-related delamination growth in a fatigue specimen with a through-width delamination. To perform the analysis, an approximate superposition analysis was developed. The analysis expresses G_I and G_{II} in closed form, which can be used easily to determine the effects of various parameters. The analysis agreed very well with more rigorous solutions.

The response of the delaminated laminate to applied loads was found to be very complex. Key observations are:

1. G_{II} is generally much larger than G_I.
2. G_I and G_{II} usually reach their peak magnitudes at different points in a fatigue cycle. G_{II} always reaches its peak value at maximum load.
3. High delamination growth rates were accompanied by large values of G_I.
4. Slow growth rates were observed even when G_I was negligibly small. This growth apparently was due to a large value of G_{II}.

APPENDIX

A strength of materials analysis is described herein for the configuration in Fig. 1.

The configuration is subdivided into four regions, as shown in Fig. 2. Because of symmetry, only half of the laminate is modeled. The laminate is of width b. The following assumptions are made:

1. Regions B and C are perfectly bonded. Regions A and D are unbonded.
2. Regions A, B, and C have constant axial strain. Hence, the force-displacement relationships are those for a simple rod subjected to axial load.
3. Region D has zero slope at both ends.
4. Region D has an initial sinusoidal imperfection of peak magnitude δ_0. The initial shape is given by

$$v(x) \mid_{\text{initial}} = \frac{\delta_0}{2} \left(1 - \cos \frac{\pi x}{a} \right) \tag{14}$$

where $v(x)$ = the distortion in the y direction.

To describe the nonlinear behavior of Region D, Eqs 15 [5] and 16 [6] for post-buckling of a column were used.

$$P_D = \frac{\pi^2 D}{a^2} \frac{\delta}{\delta + \delta_0} \tag{15}$$

$$a - \bar{a} = \frac{\pi^2}{16a} (\delta^2 + 2\delta\delta_0) + \frac{aP_D}{S_D} \tag{16}$$

where δ, a, $\bar{a}$, and P_D are peak lateral deflection, axial length before deformation, axial length after deformation, and load, respectively. Equations 15 and 16 were derived using strength of materials analysis of a column.

To combine Regions A, B, C, and D, equilibrium and compatibility conditions must be considered. The equilibrium condition for the axial force is

$$P_A + P_D = P_B + P_C = P_T \tag{17}$$

Compatibility requires the shortening of Regions A and D to be identical. Hence

$$\frac{P_A a}{S_A} = a - \bar{a} \tag{18}$$

Equations 15 through 18 can be combined to obtain the governing equation for the laminate.

$$P_T = \frac{\pi^2 S_A}{a^2} \left[\frac{1}{16}(\delta^2 + 2\delta\delta_0) + \frac{D(S_A + S_D)}{S_A S_D} \frac{\delta}{\delta + \delta_0} \right] \tag{19}$$

For a specified load, P_T, Eq 19 is solved using a Newton-Raphson technique to obtain δ. P_D can then be calculated using Eq 15. From static equilibrium, the moment acting on the delaminated region at the crack tip is

$$M = \frac{P_D}{2}(\delta + \delta_0) = \frac{\pi^2 D}{2a^2} \delta \tag{20}$$

The force, P_C, is found from rule of mixtures as

$$P_C = \frac{S_D}{S_A + S_D} P_T \tag{21}$$

References

[1] Whitcomb, J. D., *Journal of Composite Materials*, Vol. 15, Sept. 1981, pp. 403–426.
[2] Rybicki, E. F. and Kanninen, M. F., *Engineering Fracture Mechanics*, Vol. 9, No. 4, 1977, pp. 931–938.
[3] Ramkumar, R. L., "Performance of a Quantitative Study of Instability-Related Delamination Growth," NASA CR-166046, National Aeronautics and Space Administration, March 1983.
[4] Wilkins, D. J., "A Comparison of the Delamination and Environmental Resistance of a Graphite-Epoxy and a Graphite-Bismaleimide," Naval Air Systems Command Report NAV-GD-0037, 1981.
[5] Brush, D. O. and Almroth, B. L. in *Buckling of Bars, Plates, and Shells*, McGraw-Hill Book Co., Inc., New York, 1975, pp. 13–14.
[6] Ashizawa, M., "Fast Interlaminar Fracture of a Compressively Loaded Composite Containing a Defect," presented at the Fifth DOD/NASA Conference on Fibrous Composites in Structural Design, New Orleans, La, Jan. 1981; Douglas Paper No. 6994.

Alton L. Highsmith,[1] *Wayne W. Stinchcomb,*[1] *and Kenneth L. Reifsnider*[1]

Effect of Fatigue-Induced Defects on the Residual Response of Composite Laminates

REFERENCE: Highsmith, A. L., Stinchcomb, W. W., and Reifsnider, K. L., **"Effect of Fatigue-Induced Defects on the Residual Response of Composite Laminates,"** *Effects of Defects in Composite Materials, ASTM STP 836,* American Society for Testing and Materials, 1984, pp. 194–216.

ABSTRACT: This paper is concerned with the understanding, interpretation, and engineering representation of fatigue damage development in continuous-fiber composite laminates that are fiber dominated, but have significant effects caused by damage in the off-axis plies. It is the objective of this paper to present a description of damage development under tension-tension, tension-compression, and compression-compression cyclic loading that is as complete and generic as possible in the sense that it can be used as a basis for engineering representations and interpretations of the effect of such damage on the strength, stiffness, and life of composite laminates.

KEY WORDS: composite materials, laminates, fatigue (materials), fracture mechanics, localization process, transverse (matrix) cracking, characteristic damage state, fiber fracture, delamination, stiffness, stress redistribution

We are concerned with continuous, high-modulus fiber composite laminates manufactured by lamination of various plies at several angles to the direction of subsequent uniaxial loading, in the manner that has become familiar and popular for engineering purposes. Loading is assumed to be cyclic (fatigue) with constant stress amplitude for a given load history but with various combinations of tension and compression load excursion amplitudes from test to test. A special effort is made to identify generic aspects of damage development under such circumstances. Much of the data quoted was developed from "identical" material during an investigation that included careful damage monitoring for several different *R* ratios. However, the choice of "generic" behavior was made in the context of our experience over the last 14 years in this field and with an awareness of the extensive experience reported in the literature.

[1] Graduate research assistant, and professors, respectively, Engineering Science and Mechanics, Virginia Polytechnic Institute and State University, Blacksburg, Va. 24061.

One of the most complete recent directed descriptions of damage development appears in Ref [1],[2] a book which resulted from an American Society for Testing and Materials (ASTM) Symposium on Damage in Composite Materials. One of the conclusions mentioned in the summary of that book was that "experimental methods have developed to the point where nearly any type of damage can be detected." But it was also mentioned that a unified philosophy, especially a functional general engineering approach to representation has not yet been found. Among the more recent additions to the literature are a discussion of damage mechanisms observed in tension-tension loading [2,3], a discussion of the mechanics of fatigue development [4], and a discussion of the fatigue mechanisms that reduce strength, stiffness, and life for low-stress-level long-life tests in a companion paper to this one by Jamison et al in this volume. Damage accumulation and development is also discussed by Poursartip, Asby, and Beaumont, by Valentin and Bunsell, and by O'Brien (in separate papers) in Ref [5]. On the more basic side, important new groundwork is being laid in the development of viscoelastic models for damage development around individual defects by Schapery [6,7].

It is difficult to find a general quantitative characterization of fatigue damage development in composite laminates under arbitrary load excursions with tensile and compressive components. However, it is the basic premise of this paper that such a philosophy can be built around the single concept that fatigue damage development in such laminates can be generalized as a progressive localization of damage, at least in those instances when degradation is defined by loss of strength, and "failure" is defined as specimen fracture. In this sense, one might define fatigue damage development as progressive localization and contrast it to damage development under quasi-static loading that might be called spontaneous localization.

Localization—The Process

During quasi-static loading to fracture, the region of separation generally involves a localization of damage that is driven by stored strain energy in a region of damage initiation. In that sense, the localization could be called spontaneous. As an example, one might consider fracture under monotonic tensile loading for a quasi-isotropic laminate. As we noted in earlier papers [2,3,4], the localization can be described in an approximate way by matrix cracking that causes *local* net-section stresses in the 0-deg plies to be elevated by the load "dropped" by the off-axis plies that crack, leading to overload and fracture. When laminate strengths are calculated using such assumptions (commonly called the discount scheme), the quasi-static strength of composite laminates (not dominated by edge-related damage) can be predicted with engineering accuracy. Such calculations are the "accepted method" and have been used for over 15

---

[2] The italic numbers in brackets refer to the list of references appended to this paper.

years by most engineers and investigators in the field. However, under cyclic loading, two major aspects of the problem are different. First, the cyclic load amplitude is lower than the static ultimate strength. Hence, the localization process must be more severe to produce fracture at such a stress level. And second, the localization is a progressive process that is driven not only by stored strain energy but also by the cyclic energy dissipated (the cyclic work rate) as the applied load oscillates. Hence, damage development is progressive, including such behavior as delamination growth, crack coupling, etc. The localization process under such conditions is also progressive, and must (ultimately) reduce the strength of the laminate to the applied stress level if fracture occurs.

It will be convenient for us to discuss progressive localization as a process having two distinctive features that form the basis for a discussion of the mechanics of the process. Those features are local changes in geometry and local changes in stress. Local changes in geometry are caused by such things as crack formation, interface separation, and fiber fracture. When such micro-events cause local geometry changes, the local stress state also changes, that is, a cracked ply "shifts" its load to neighboring plies in the region near a crack, stress concentrations may form, etc. While these two features generally occur concomitantly, we will use them as discrete features that occur in an alternating chain of events that (collectively) cause progressive localization, strength reductions, and eventual fracture if cycling continues. Our detailed discussions that follow will be arranged in such a sequence for each generic case. This step-by-step analysis is motivated by observations that suggest that damage commonly develops in characteristic patterns (which we have called "characteristic damage states" (CDS) in earlier publications) that can be modeled with well-set mechanics formulations and used to anticipate both the consequences of such patterns (reduction in strength and stiffness) and the next sequence of damage events to follow under continued cyclic loading. As the next sequence of damage develops, the local geometry changes to form a new pattern and new local stress changes, producing another "link in the chain." In this way, the chain of events that cause progressive localization are described and analyzed. It is apparent that this approach is very convenient for the eventual use of such information to construct a cumulative damage philosophy.

Tension-Tension Loading

Before discussing the effects of fatigue-induced defects on the response of composite laminates, it is useful to determine what the "critical elements" are for the laminate and loading type under consideration. As mentioned previously, if fatigue failure is defined by specimen fracture and if the laminate under consideration contains 0-deg plies, then the 0-deg plies are the critical elements when the laminate is subjected to tension-tension fatigue loading, since 0-deg ply failure and specimen fracture are coincident. The off-axis plies then constitute the "sub-critical elements" of the laminate.

The identification of critical and sub-critical elements provides a framework for a discussion of the effects of particular damage modes on residual response. An overview of the localization process for tension-tension fatigue is presented in Table 1. In analytical terms, a damage event represents a change in geometry (boundary conditions) within a laminate. Such a change in geometry changes the local stress distribution. In the discussion that follows, it will become apparent that a given damage mode may influence laminate residual response in two ways. First, the stress redistribution associated with the damage may involve the critical elements, in this case, the 0-deg plies, directly. If damage causes a local increase in the fiber direction stress in the 0-deg plies, a corresponding decrease in residual strength is expected. Second, the stress redistribution associated with the damage may encourage further damage development.

As indicated in Table 1, the first damage mode observed in a composite laminate subjected to tension-tension fatigue loading is matrix cracking in off-axis plies. A matrix crack extends completely through the thickness of the cracked ply and follows the general fiber direction of the ply across the width of the specimen. Since the crack surfaces are stress free, there is a volume of material surrounding the crack in the cracked ply that carries only a fraction of the load that is carried by that ply in an undamaged region. This means, in turn, that the adjacent uncracked plies must carry an increased load. Figure 1 is a moiré fringe

TABLE 1—*Process of localization for tension-tension fatigue loading.*

Activity In:			
Critical Element		Subcritical Element	
Geometry Change	Stress State Change	Geometry Change	Stress State Change
		Matrix crackings CDS formation	
	Increase of σ_x in 0 degree plies, local stress concentration		Local variation of σ_{ij} in θ plies
Local fiber failures near matrix cracks			
	Stress concentrations near fiber failures		
		Fiber direction cross-cracking	
	Increase interface stresses		Increase interface stresses
		Local delamination	
	Change of all local stresses		Change of all local stresses
Initiation event			
	Local overload or critical stress localization		
Specimen fracture			

Increasing cycles (time)

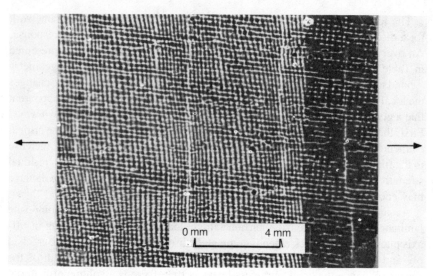

FIG. 1—*Moiré fringe pattern of a* [0,90₃]ₛ *E-glass laminate between cracked sections in the 90-deg plies.*

pattern taken from the 0-deg ply of a $[0,90_3]_s$ E-glass specimen that was loaded quasi-statically to develop matrix cracks in the 90-deg plies [8]. Each dark or light band is a contour of constant relative axial displacement. Consecutive dark or light bands differ in displacement by 1/1200 mm. Near cracked sections of the interior 90-deg plies, the observed fringe density, and thus the axial strain in the 0-deg plies, is high. In regions between matrix cracks, the fringe density, and thus the 0-deg ply strain, is low. At the cracked sections, load that would normally be carried by the 90-deg plies must be carried by the 0-deg plies. Also, note that the axial strain in the 0-deg plies is not uniform because stress is transferred back into the 90-deg plies away from the crack. Thus, the axial strain in the 0-deg plies is largest at the cracked cross sections, and decreases with increasing distance from the cracked sections.

During cyclic loading, numerous matrix cracks develop in the off-axis plies. They do not develop randomly, but form a characteristic pattern called the Characteristic Damage State (CDS) that depends only on material properties, ply thicknesses, and stacking sequence [9,10]. After sufficient cyclic loading, the cracks within each of the off-axis plies achieve a uniform spacing that is determined by the ability of the surrounding plies to transfer load back into the cracked ply. A simple example of a CDS is shown schematically in Fig. 2. Material properties and ply constraint influence the distance over which stress recovery in the cracked ply is complete. It is the redistribution of stresses near matrix cracks that determines the CDS.

Because matrix cracks develop when a laminate is loaded quasi-statically, they are not expected to influence directly the residual strength of a laminate

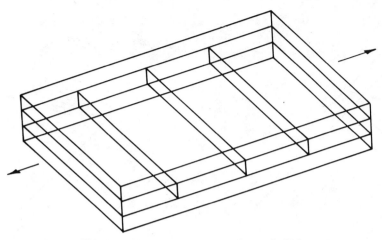

FIG. 2—*Schematic representation of a simple CDS.*

when they are developed under tensile fatigue loading. However, because they disturb the internal stress distribution, they influence residual response by encouraging additional damage development. Further, this damage tends to be localized: since the local stresses near matrix cracks motivate damage development, the damage develops in the vicinity of the matrix cracks.

The axial stress in a 0-deg ply is elevated near a transverse crack in an adjacent ply since the stress in the adjacent ply is zero at the crack surface. The 0-deg ply carries some of the load that would be carried by the adjacent ply if it were undamaged. If local deformation is constrained by surrounding material, the stress in the 0-deg ply is also elevated by the crack as a stress raiser, that is, a stress concentration. These high stress regions in the 0-deg plies are regions that are favorable for 0-deg fiber fracture. Figure 3b shows schematically how fiber fractures are clustered near a transverse crack in an adjacent ply. Scanning electron microscope (SEM) micrographs of a 0-deg ply taken from a fatigue damaged $[0,90_2]_s$ T300/5208 graphite/epoxy specimen are shown in Fig. 4 [3]. In the first micrograph, numerous fiber fractures are evident. Also, note the dust-like traces of gold chloride that are observable in this region. The gold chloride was dissolved in diethyl ether and applied to the specimen edge. Traces of the gold chloride were left along matrix crack lines adjacent to the 0-deg ply. Note that the fiber fractures observed in Fig. 4a occur only in the region marked by the gold chloride, and that in Fig. 4b, where there are no traces of gold chloride, there are no fiber fractures. Jamison [11] has found that the number of fiber fractures in the interior of a given 0-deg ply is smaller than the number of fiber fractures near a ply interface, as the effect of the transverse cracks is diminished. In addition, the zone of fiber cracks is wider in the interior of the ply. As indicated in Fig. 5, most of these fiber fractures occur relatively early in the fatigue life of a laminate.

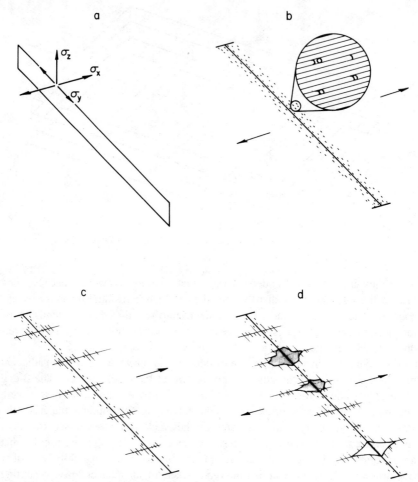

FIG. 3—*Schematic representation of the progressive development of* (a) *matrix cracks in off-axis plies,* (b) *fiber fractures,* (c) *fiber direction cross cracks, and* (d) *local delamination.*

Fiber fractures in the 0-deg plies affect laminate response in two ways. First, a fiber break causes a local increase in the axial stress in the 0-deg ply, which is the critical element in tension-tension fatigue, in much the same manner as a transverse crack in an adjacent ply does. An analysis by Hedgepeth and Van Dyke [12] indicates that this effect is restricted to a zone of a few fiber diameters surrounding the crack. Another similarity to the cracked adjacent ply is that stress is transferred back into the fiber away from the crack by its surroundings. As indicated in a shear lag analysis by Rosen [13], significant shear stresses develop in the matrix material surrounding the broken fiber in order to cause this stress transfer. Since fiber breaks are most common along matrix cracks in

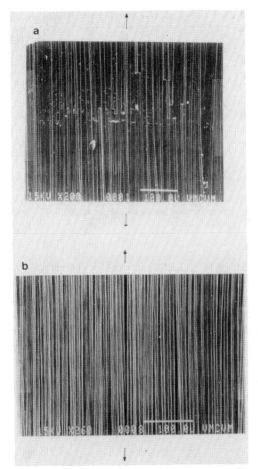

FIG. 4—*Scanning electron micrographs taken from a fatigue-damaged* $[0,90_2]_s$ *graphite/epoxy specimen showing* (a) *fiber fracture zone, and* (b) *a fiber fracture-free zone.*

adjacent plies, this elevated shear stress contributes to the local degradation of the interface.

The next damage mode that is observed is fiber direction cross cracking that develops at ply interfaces [3,11]. A representative damage state containing fiber direction cross cracking is shown schematically in Fig. 3c. A penetrant-enhanced X-ray radiograph of a $[0,90,\pm45]_s$ graphite/epoxy specimen that was subjected to tension-tension fatigue is presented in Fig. 6. Note the numerous small dark lines, which are penetrant enhanced cracks, extending in the 0-deg direction. An examination of stereo X-ray pairs reveals that most of these cracks do not extend through the thickness of the 0-deg ply initially, but are restricted to the ply interface, possibly to the matrix-rich interfacial region between the 0-deg

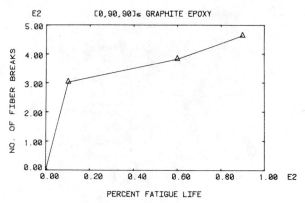

FIG. 5—*Chronology of fiber fracture during the fatigue life of* $[0,90_2]_s$ *graphite/epoxy laminates (adapted from Ref. 11).*

and 90-deg plies. There is, in fact, a network of such cracks parallel to both the 0-deg and 90-deg direction at the 0-90 interface associated with each transverse crack in the off-axis ply. Some of these microcracks may eventually propagate through the thickness of the 0-deg ply.

Referring to Table 1, both matrix cracking and fiber fracture influence cross-crack development. The elevated shear stress in the matrix near fiber fractures at the interface contributes to the overall degradation of the interface near a transverse crack, but the greatest contribution is made by the transverse crack itself. Generally, the transverse crack in the cracked ply serves as a stress raiser. The stresses due to the crack dominate the local response, and, as a consequence,

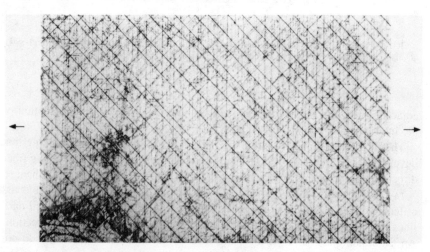

FIG. 6—*Section of a radiograph showing fiber direction cross cracking in a* $[0,90,\pm45]_s$ *graphite/epoxy laminate.*

there are large tensile stresses in the uncracked ply normal to the plane of the crack (shown as σ_x in Fig. 3a). If the uncracked ply were not constrained along the crack line by the cracked ply, a large Poisson contraction would be observed. But in fact, since the crack line is in the fiber direction of the cracked ply, it exerts maximum constraint against just such a contraction. The result is that a large tensile stress is developed in the uncracked ply at and near the interface between the two plies in the direction of the crack line (shown as σ_y in Fig. 3a). This tensile stress encourages the formation of fiber direction cross cracks.

Presently, the influence of fiber direction cross cracks on the critical elements and their direct influence on laminate residual response is unclear. They do, however, appear to encourage the development of local (internal) delaminations. A radiograph of a fatigue-damaged $[0,90_2]_s$ graphite/epoxy specimen is presented in Fig. 7. In this laminate, in which both the global and local transverse stress in the 0-deg plies are tensile, cross cracks in the 0-deg plies grow through the thickness to become longitudinal splits. The gray zones that extend generally in the longitudinal direction are local delamination of the 0-90 deg interface. Notice that each such zone is associated with a particular longitudinal split. A schematic representation of this damage state is shown in Fig. 3d. These delaminations develop because of interlaminar stresses that arise due to the presence of trans-verse matrix cracks, longitudinal splits, and fiber direction cross cracks in the laminate. When load is applied to the specimen, these cracks open up. The constraining plies resist this deformation via tensile interlaminar normal stress, σ_z, and interlaminar shear stress τ_{xz}. Talug [14] has developed a three-dimensional finite difference analysis that can compute three-dimensional field stresses near matrix cracks in a laminate. Figure 8 shows a plot of σ_z and τ_{xz} through the

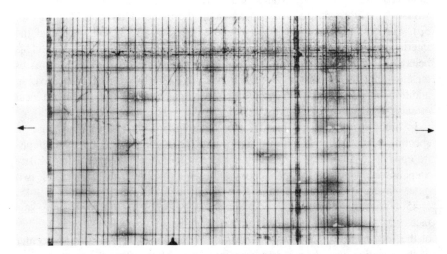

FIG. 7—*Section of radiograph showing local delamination associated with longitudinal splitting in $[0,90_2]_s$ graphite/epoxy laminates.*

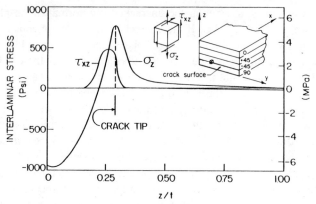

FIG. 8—*Distribution of interlaminar stresses σ_z and τ_{xz} near a crack in the 90-deg plies of a $[0,\pm45,90]_s$ graphite/epoxy laminate. (A half-thickness edge view is shown.)*

thickness of a $[0,\pm45,90]_s$ graphite/epoxy laminate near a crack in the 90-deg plies. Near the crack tip, σ_z has a relatively large tensile value, and τ_{xz} is also large. Both of these stresses tend to cause the plies to delaminate near the crack. Both matrix cracks and fiber direction cross cracks produce these types of stresses. While matrix cracks cause greater redistributions of stresses, cross cracks are typically more numerous and seem to be a precursor of local delamination.

Here a comment should be made concerning the effect of matrix cracks and cross cracks on the local distribution of stress in the laminate. Consider the complete stress field near a matrix crack to be the superposition of a global (laminate) stress field and a local stress field. Upon isolating the local stress field near the crack, one would find that in the adjacent plies, three normal stresses [the normal stress perpendicular to the plane of the crack (σ_x in Fig. 3a), the normal stress along the line of the crack (σ_x in Fig. 3a), and the interlaminar normal stress (σ_z in Fig. 3a)] all take on large tensile values. This behavior is generic, since the only condition necessary to render these stresses tensile—that the cracks lie along the fiber direction—is always observed. In addition, while in principle it is possible that global stresses dominate these local effects, we have observed no such behavior.

In order to investigate the effect of delamination on laminate response, a specimen with a delamination was fabricated. Three pieces of T300/5208 graphite/epoxy, 254-mm long by 38.1-mm wide, were bonded together using Miller-Stephenson Type 907 two-part epoxy adhesive. As shown in Fig. 9a, the two outer layers were $[0,90]_s$ material, while the central layer was $[+45, -45, -45, +45]_s$ material. A 50.8-mm-long region on each of the mating surfaces was coated with vaseline before the adhesive was applied. During assembly of the specimen, a 38.1-mm-wide strip of mylar film was placed on the center of the vaseline-coated region of each interface. After curing the adhesive, the mylar strips were removed, and the laminate was trimmed to 25.4 mm in width.

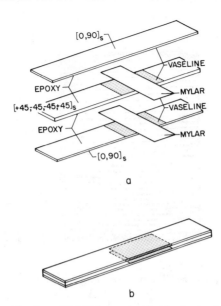

FIG. 9—*Schematics of Specimen SC-1 (fabricated with delaminations)* (a) *before and* (b) *after assembly.*

The resulting specimen, shown in Fig. 9b, contained 50.8-mm-long delaminations in the 0-45 interfaces that crossed the entire width of the specimen and were aligned along its length.

A moiré fringe pattern, Fig. 10, obtained from this specimen indicates how the delamination influences material response. Again, each fringe is a contour of relative axial displacement. In the undamaged region, the fringes are straight and uniformly spaced. In the delaminated region the fringes are curved and are also closer together than those in the undamaged region. The difference in fringe density indicates that the axial strain is greater in the delaminated region. This difference in strain arises because the three layers are not constrained to have the same transverse and in-plane shear strains in the delaminated region. The Poisson's ratio mismatch between the two types of layers causes the distributions of stresses in the two regions to differ. This also explains why the moiré fringes are curved in the delaminated region. The transverse constraint relaxation takes place more readily near the specimen edge, and thus the transition from undamaged to delaminated response occurs over a shorter distance. Away from the edge near the delamination front, the delaminated layers are still somewhat constrained by the surrounding material. A simple stress analysis using laminate theory can be performed to estimate the effect of delamination on internal stresses. In this analysis, the three layers in the delaminated region are required to have equivalent longitudinal strain, but no restrictions are imposed on the transverse and shear strains. Equilibrium requires that the delaminated and undelaminated regions are subjected to the same total load. The predictions of in-plane stresses

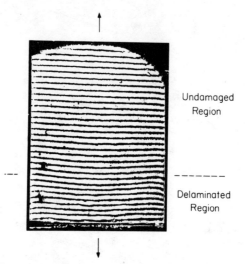

Undamaged
Region

Delaminated
Region

FIG. 10—*Moiré fringe pattern taken from Specimen SC-1 spanning both undamaged and delaminated regions.*

in each ply of both the delaminated and undamaged regions are presented in Table 2. Note that σ_x in the 0-deg plies is larger in the delaminated region, as indicated by the moiré pattern. Using a 1-in.-gage-length clip-on type extensometer, local axial strain response was monitored in both the undamaged and delaminated regions. This data is presented in Fig. 11. (Load was chosen instead of stress because the material cross sections in the delaminated and undamaged regions differ by two thicknesses of epoxy adhesive. However, the adhesive is relatively compliant so this discrepancy should not significantly affect the results.) The reduction of axial stiffness due to the delamination was observed to

TABLE 2—*Comparison of in-plane stresses predicted via laminate analysis for a*
$[0,90,90,0, +45, -45, -45, +45]_s$ *graphite/epoxy laminate (lamina properties:* $E_1 = 141.6$
GPa, $E_2 = 10.2$ *GPa,* $\nu_{12} = 0.305$, $G_{12} = 6.00$ *GPa,* $N_x = 113.0$ *kN-mm) with and without delamination at the 0-45 interfaces.*

Ply	Damage Condition	σ_x, MPa	σ_y, MPa	τ_{xy}, MPa
0	undamaged	399.4	0.1	0.0
	delaminated	459.1	8.8	0.0
90	undamaged	26.3	−112.6	0.0
	delaminated	32.7	−8.8	0.0
+45	undamaged	100.3	56.2	65.1
	delaminated	67.2	0.0	27.9
−45	undamaged	100.3	56.2	−65.1
	delaminated	67.2	0.0	−27.9

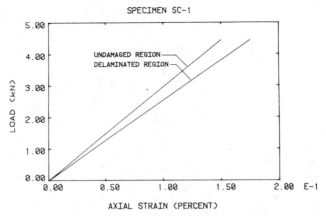

FIG. 11—*Comparison of load-strain response of undamaged and delaminated regions of Specimen SC-1.*

be 14.3%. The simple analysis just discussed predicts a 12.7% reduction in stiffness.

It is true that the delaminations in the specimen just discussed can hardly be considered local because they cover a large portion of the adhesive interfaces. Local delaminations appear to be more constrained by the surrounding material. However, interaction of other damage events with the local delaminations have thus far been ignored. Quasi-static tension tests were performed on $[0,90,\pm45]_s$ graphite/epoxy specimens that had aligned delaminations at both 0-90 interfaces. These delaminations were produced by taping together two 25.4-mm-wide strips of mylar fiber stacked atop one another, and placing one such delamination at each 0-90 interface during the layup of the laminate prior to curing. During tensile loading, axial strain was monitored via both a 25.4-mm-gage-length clip gage that spanned just the delaminated zone, and a 50.8-mm clip gage that spanned both the delaminated region and undamaged material. While no measurable difference in stiffness was detected (or predicted), the average reduction in tensile strength from that observed in specimens without delaminations was 25.4%. This is nearly the reduction in strength that would be expected if the interior plies carried no load. It is possible that other damage is coupling with the delamination and mechanically separating the 0-deg plies from the interior plies.

At this point in the history of the specimen or component, most of the "fatigue effect" has occurred. The cumulative damage state to this point is shown schematically in Fig. 12. In the context of our present terminology, the development of damage in subcritical elements has redistributed the load carried by various plies and has caused a progressive localization of deformation and stress that has increased the stress in the remaining critical elements that are yet unbroken. We have seen that the local stress amplitudes that can be produced by this process,

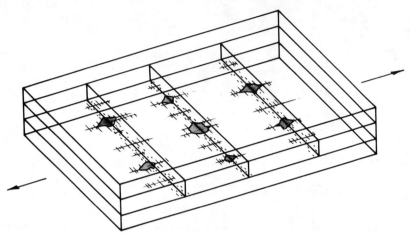

FIG. 12—*Schematic representation of the cumulative damage state through the development of local delaminations.*

especially by the newly discovered local delamination mode, are sufficient to cause fracture of critical elements such as 0-deg plies at applied stress levels that are as low as 50 to 60% of the quasi-static loading strength of an undamaged laminate, in general terms.

The next stage in the life of the specimen or component is a geometry change referred to as the "initiation event" in Table 1. At present, this event has not been isolated. From our experience in the laboratory, previously outlined, we can say that the local conditions may be sufficiently severe to cause fracture, but the event that actually causes the final fracture initiation has not been identified despite considerable effort to do so. The data we have obtained is at least consistent with one scenario that is also suggested by various analytical treatments and some experimental results reported in the literature. That scenario suggests that the critical initiation event is the fracture of a critical number of adjacent fibers, for example, that for a given final ply fracture to occur in a given material system, a critical number of fibers must break in a region small enough so that the fractures "couple up" to form a local region that reaches a critical size (characteristic of that material) that is just large enough to cause cataclysmic propagation and specimen fracture. This scenario is supported by the fiber failure data observed in our laboratory in the sense that the number of fibers that fail in close proximity to one another increases throughout the fatigue test and appears to reach a maximum number that is surprisingly small and surprisingly consistent. Tamuzs has made similar observations and has used a penny-shaped crack, fracture mechanics analysis to predict such behavior with some success [15]. Batdorf has generated a statistical strength argument that is based on a very similar physical concept [16]. In both cases, the critical number of broken fibers is similar to our experience for graphite epoxy, generally of the order of half a dozen or less. Since no specific determination of such a number was attempted in the present investigation (such a determination would require an extensive data base meticulously generated and interpreted for this specific purpose), no specific numbers are quoted here. It should also be mentioned that the statistical interpretations of strength derived by Phoenix and Harlow are also consistent with this "critical fiber fracture localization" concept [17]. The authors know of no evidence that such a concept could *not* be the basis for fracture initiation.

However, we must stop short of a conclusion on this matter until positive and direct identification has been made. Returning to Table 1, we note that when the local geometry is altered by the fracture initiation event, the corresponding stress-state change is expected to be either local overload or other type of critical stress localization in the critical element. It may be that a critical rate of strain energy release is reached, or other criteria may be more appropriate. In any case, a comparison of our observations of specimens that were severely damaged so that their residual strength was reduced almost to the level of applied stress, and our observations of specimens after fracture following long-life cyclic loading has revealed no damage modes or local geometric details that suggest that a new type of event initiates fracture. In fact, our experience to date does not contradict the possibility that the fracture initiation event is the same in severely fatigue-damaged specimens and specimens loaded to failure quasi-statically. It could be, for example, that the same number of local adjacent fiber breaks trigger the fracture event in both cases. While such a suggestion is not contradicted by our findings, it is certainly not established by them either. However, such a conclusion would be of great value to efforts to model residual strength and life, so that further work in this area is encouraged by the authors. It should also be remembered that the final fracture event may be peculiar to a given composite material system, and is certainly limited (in the present case) to fiber-dominated fracture in tensile loading.

Compression-Compression Loading

In sharp contrast to the extensive documentation of the details of damage due to tension-tension loading, relatively little is known about damage mechanisms and subsequent response of laminates subjected to compression-compression loading. Much of the compression data that exists must be viewed in the light of methods of test and special fixtures designed to restrict or prevent buckling. Thus, compression data may represent the behavior of specimens subjected to particular test procedures and may not be representative of the compression response of the material. The data presented in this discussion were obtained from 48-ply quasi-isotropic laminates of AS-3502 graphite epoxy [18]. Three stacking sequences were investigated: (a) $([0,\pm45,90]_s)_6$, (b) $([0,90,\pm45]_s)_6$, and (c) $([0,+45,90,-45]_s)_6$.

Fatigue tests were conducted at a stress ratio of $R = 10$ at 10 Hz on specimens

2.54-cm wide with a free, unsupported length of 10.2 cm. Under compressive loading, the stress perpendicular to the fibers in the 90-deg plies is compressive and matrix cracks do not initiate in these plies. The primary damage mode under compression-compression loading is delamination of ply interfaces where the interlaminar normal stresses (σ_z) are tensile or the interlaminar shear stresses are large or both. Of the three stacking sequences investigated, only the ($[0,90,\pm45]_s)_6$ laminate exhibited compression-compression fatigue damage at minimum cyclic strains of -5000, -5500, and -6000 $\mu\epsilon$. The fatigue lives corresponding to these cyclic strains are approximately 10^4, 10^5, and 6×10^5 cycles, respectively. The damage in these cases was edge delamination of the ±45 interfaces where the interlaminar normal stresses are tensile and the interlaminar shear stresses are large. At minimum cyclic strains of -4500 $\mu\epsilon$, neither delaminations nor matrix cracks developed and the specimens survived one million cycles. The other two laminates survived one million cycles with no damage at minimum cyclic strain between -4500 and -6000 $\mu\epsilon$.

When delamination is the primary damage mode of laminates subjected to compression-compression loading, the critical element is defined by the stiffness of the various delaminated regions and the loads carried by those regions. If, for example, the outer 0, 90, and $+45$-deg plies of the ($[0,90,\pm45]_s)_6$ laminate are separated from the rest of the laminate due to delamination of the first ±45 interface, the three delaminated plies will deform out-of-the plane due to a decrease in their bending stiffness as the delamination grows along the length and through the width of the specimen.

The increased load carried by the core region of the laminate and the local stiffness reduction of the core resulting from the delamination promote the development of additional delaminations throughout the thickness of the laminate until the combination of locally high loads and reduced stiffness causes the laminate to buckle.

The scenario of the development of damage in laboratory specimens and structural components will vary according to a number of geometrical factors. The fatigue life of a specimen is dependent on the growth rate of delaminations, which, in turn, is a function of the unsupported length of the specimen and the mode of gripping. A delamination that causes a compressive instability of a laboratory specimen may have little or no effect on the response of a structure. The effects of defects on compressive fatigue life of a laminate must be understood to be a combined effect of material, geometrical, and structural properties.

Tension-Compression Loading

Having discussed the fatigue response of laminates under tension-tension and compression-compression, one might assume that the tension-compression response could be described as a simple combination of damage due to these two loading modes. This is not the case. There is a complex interaction between tensile damage modes and compressive damage modes that Whitney has modeled as competing failure modes [19].

Tension-compression fatigue tests at $R = -1$ have been conducted on the three 48-ply quasi-isotropic laminates described in the previous section and the data are reported in Ref 18. The history of damage development can be summarized in the following way. During the tensile portion of the loading waveform, damage initiates in the form of matrix cracks as described in a prior section. The matrix cracks grow into the ply interfaces and couple along those interfaces, producing interlaminar shear stresses, τ_{xz}. Under tension-tension loading at $R = 0.1$ to a maximum cyclic strain, ϵ_{max}, the range of the cyclic interlaminar stresses is 0.9 $\tau_{xz\ max}$. However, under tension-compression loading between $\pm\epsilon_{max}$, the range of the cyclic interlaminar shear stresses is $2\tau_{xz\ max}$; and the interfacial damage propagates at a higher rate than it does under tension-tension loading. This response was observed in each of the three laminates under tension-compression loading even though the response of two of the laminates was insensitive to compression-compression loading. The accelerated degradation of the laminates, shown in Figs. 13 through 15, is due to the cyclic nature of the stress fields associated with the state of damage. When damage is present, the associated stresses cause damage propagation on both the tensile and compressive portions of the loading waveform. This observation is supported by experimental data on the fatigue response of the $([0,\pm45,90]_s)_6$ graphite/epoxy laminate. Under tension-tension loading at a maximum strain of 4500 $\mu\epsilon$ and $R = 0.1$, matrix cracks in off-axis plies initiate early in the loading history, as shown in Fig. 13.

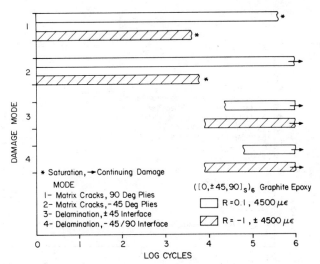

FIG. 13—Chronology of damage modes in $([0,\pm45,90]_s)_6$ graphite/epoxy laminates for cyclic loading at R = 0.1 and −1.

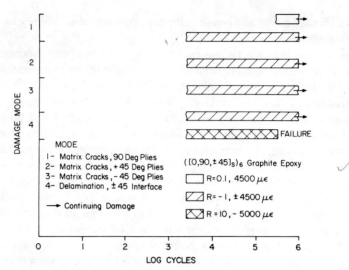

FIG. 14—*Chronology of damage modes in* $([0,90,\pm45]_s)_6$ *graphite/epoxy laminates for cyclic loading at* $R = 0.1, -1,$ *and 10.*

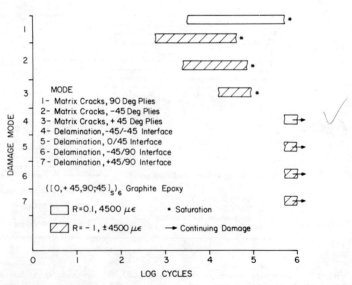

FIG. 15—*Chronology of damage modes in* $([0,+45,90,-45]_s)_6$ *graphite/epoxy laminates for cyclic loading at* $R = 0.1$ *and* $-1.$

Under compression-compression loading at a minimum strain of -4500 $\mu\epsilon$ and $R = 10$, no damage was observed after one million cycles. In a subsequent experiment to investigate the effects of a loading history with tensile and compressive cycles, a specimen of this laminate configuration was subjected to an initial block of tension-tension loading for 1000 cycles. Nondestructive inspection showed the expected matrix cracks in the off-axis plies. A second block of compression-compression loading was then applied to the damaged laminate. After 1000 cycles, the nondestructive inspection showed that the cracks that initiated during the block of tensile loading had coupled along ply interfaces during the block of compressive loading, even though a similar compressive loading history of an undamaged specimen did not initiate damage.

Figures 13 through 15 identify damage modes and compare the chronology of damage in the three laminates under tension-tension loading ($R = 0.1$, $\epsilon_{max} = 4500$ $\mu\epsilon$) and tension-compression loading ($R = -1$, $\epsilon_{max} = 4500$ $\mu\epsilon$). In general, tension-compression loading was the more damaging of the two loading modes. Under tension-compression loading, matrix cracks initiated earlier in the loading history, cracks reached saturation spacing in some laminates earlier in the loading history, and more extensive delamination was observed. Under compression-compression loading ($R = 10$, $\epsilon_{min} = -4500$ $\mu\epsilon$), no damage was detected in any of these laminates. These results suggest that the damage modes in laminates subjected to tension-compression loading are both interactive and competitive. The effects of these damage modes on the response of the laminates are most detrimental for tension-compression loading.

Summary

We have seen that damage development under fatigue loading can be regarded as a process that usually involves sequential changes in geometry and load distribution at the local level. We have also seen that this process is very distinctive for tension-tension, compression-compression, and tension-compression loading. The concept of progressive localization has been advanced as a general definition of the fatigue process. In the case when failure is defined by fracture, that is, separation of the specimen into two pieces, the progressive localization generates a local region of damage that eventually becomes the fracture site. In the case of failure by instability, the localization process acts to disconnect various elements of the laminate until the resistance to buckling is lowered to the applied stress level.

The concept of critical and subcritical elements was also introduced. Fatigue damage, *per se*, was defined as the degradation of subcritical elements. Resistance to failure (residual strength) was defined in terms of the remaining integrity of the critical elements and the localized stress field caused by subcritical element damage.

As mentioned in the beginning of this paper, a major motivation for the arrangement of information and the architecture of the interpretations presented in this paper was the apparent utility of both for the purpose of the description of damage accumulation in composite laminates under fatigue loading with arbitrary tensile and compressive load excursions. The authors suggest that it is possible to construct rigorous models of the local changes in geometry and redistributions of load so as to calculate the effect of the degradation of subcritical elements on the remaining strength and rate of degradation of critical elements as a function of cycles of load application, guided either by direct damage observations or a damage indicator such as stiffness change, and, thereby, to anticipate the residual strength and life of specimens or components. While it is not the charge of this paper to discuss the experience that supports such a suggestion, it is at least pertinent to stress the practical orientation of the approach we have taken to the present investigation and presentation in this paper.

Acknowledgment

The authors gratefully acknowledge the support of the U.S. Air Force under Contract No. F33615-81-K-3225 monitored by Dr. G. P. Sendeckyj, and the support of the General Dynamics Corporation under Contract No. 957229. They also express their gratitude to Barbara Wengert for typing the manuscript, to Robert Davis for assistance in specimen preparation, and to G. K. McCauley for assistance with photography.

References

[1] *Damage in Composite Materials, ASTM STP 775,* K. L. Reifsnider, Ed., American Society for Testing and Materials, 1982.

[2] Reifsnider, K. L., in *Fatigue and Creep of Composite Materials, Proceedings,* 3rd Risø International Symposium on Metallurgy and Materials Science, 6–10 Sept. 1982, Fyens Stiftsbogtrykkeri, Roskilde, Denmark, 1982, pp. 125–136.

[3] Reifsnider, K. L. and Jamison, R. D., *International Journal of Fatigue,* Oct. 1982, pp. 187–198.

[4] Reifsnider, K. L., "Damage Mechanics and NDE of Composite Laminates," *Proceedings,* IUTAM Symposium on Mechanics of Composite Materials, 16–19, Aug. 1982, Virginia Polytechnic Institute and State University, Pergamon Press, 1982.

[5] *Fatigue and Creep of Composite Materials,* H. Lilholt and R. Talreja, Eds., 3rd Risø International Symposium on Metallurgy and Materials Science, Fyens Stiftsbogtrykkeri, Roskilde, Denmark, 1982.

[6] Schapery, R. A., "Models for Damage Growth and Fracture in Nonlinear Viscoelastic Particulate Composites," MM3168-82-5, Mechanics and Materials Center, Texas A & M University, Aug. 1982.

[7] Schapery, R. A. in *1981 Advances in Aerospace Structures and Materials,* S. S. Wang and W. J. Renton, Eds., American Society of Mechanical Engineers, 1981, pp. 5–20.

[8] Highsmith, A. L. and Reifsnider, K. L., *Composites Technology Review,* Vol. 4, No. 1, Spring 1982, p. 20.

[9] Reifsnider, K. L., Henneke, E. G., and Stinchcomb, W. W., "Defect-Property Relationships in Composite Materials," AFML-TR-76-81, Air Force Materials Laboratory, June 1979.

[10] Highsmith, A. L. and Reifsnider, K. L. in *Damage in Composite Materials, ASTM STP 775,* K. L. Reifsnider, Ed., American Society for Testing and Materials, 1982, pp. 103–117.

[11] Jamison, R. D., "Advanced Fatigue Damage Development in Graphite Epoxy Laminates," PhD. thesis, College of Engineering, Virginia Polytechnic Institute and State University, Aug. 1982.

[*12*] Hedgepeth, J. and Van Dyke, P., *Journal of Composite Materials,* Vol. 1, 1967, pp. 294–309.

[*13*] Rosen, B. W., *AIAA Journal,* American Institute of Aeronautics and Astronautics, Vol. 2, No. 11, Nov. 1964, pp. 294–309.

[*14*] Reifsnider, K. L. and Talug, A., "Analysis of Stress Fields in Composite Laminates with Interior Cracks," VPI-E-78-23, Virginia Polytechnic Institute and State University, Sept. 1978.

[*15*] Tamuzs, V., "Some Peculiarities of Fracture in Heterogeneous Materials," *Proceedings,* 2nd US-USSR Symposium on Fracture of Composite Materials, Lehigh University, G. Sih and V. Tamuzs, Eds., Sythoff and Noordhoff, March 1981.

[*16*] Batdorf, S. B., "Tensile Strength of Unidirectionally Reinforced Composites—I," *Journal of Reinforced Plastics and Composites,* 1982.

[*17*] Harlow, D. G. and Phoenix, S. L., *Journal of Composite Materials,* Vol. 12, 1978, pp. 195–214.

[*18*] Liechti, K. M., Reifsnider, K. L., Stinchcomb, W. W., and Ulman, D. A., "Cumulative Damage Model for Advanced Composite Materials," AFWAL-TR-82-4094, Air Force Flight Dynamics Laboratory, Wright-Patterson Air Force Base, July 1982.

[*19*] Whitney, J. M., "A Residual Strength Degradation Model for Competing Failure Modes," presented at the ASTM Symposium on Long Term Behavior of Composites," Williamsburg, Va., 1982.

DISCUSSION

Theodore Laufenberg[1] *(written discussion)*—Recent tests run on fiberglass/wood laminates with longitudinal to transverse stiffness ratios on the order of 9:1 have yielded some information on crack density and stiffness loss for this realistic laminate. First, there is no detectable stiffness loss. The capability to detect the loss is limited by the accuracy and repeatability of the load and strain detecting equipment.

Secondly, our tracking of crack density shows, as you hypothesized, that near failure localized cracking occurs at the impending failure zone that can increase the crack density four-fold. The composite we have worked with has never reached the saturated crack density except at the local failure zone.

Alton L. Highsmith (authors' closure)—It is possible that the transverse cracks that develop in a laminate cause no appreciable stiffness loss if the broken plies are much more compliant in the load direction than the unbroken ones. However, several factors other than equipment sensitivity can affect the detection of stiffness loss. Of particular importance is the gage length of the strain measuring device. If the gage length is small, the material in the test section may not provide an accurate representation of the laminate in its damaged condition. This is especially important at low crack densities.

As far as the saturation crack density is concerned, our experience is that not all laminates reach saturation when loaded quasi-statically. Saturation crack density can be attained if the laminate is fatigued over an appropriate range of

[1] Forest Products Laboratory, Madison, Wis. 53705.

load levels. This of course implies that additional damage (for example, fiber breaks and delaminations) must develop before the laminate will fail. In laminates, where off-axis ply cracking is life-limiting, saturation crack density will not be observed. The saturation crack density is properly viewed as a limit that the crack density approaches during damage development.

Julius Jortner[1]

A Model for Predicting Thermal and Elastic Constants of Wrinkled Regions in Composite Materials

REFERENCE: Jortner, Julius, **"A Model for Predicting Thermal and Elastic Constants of Wrinkled Regions in Composite Materials,"** *Effects of Defects in Composite Materials, ASTM STP 836,* American Society for Testing and Materials, 1984, pp. 217–236.

ABSTRACT: Previous theoretical attempts to predict the mechanical behavior of curved-fiber composites have provided only upper or lower bound estimates of the elastic constants of orthotropic wrinkled regions. In this paper, a new model is described that extends the prior work in the following ways: (1) for a certain class of wrinkles the approach is exact in that stress equilibrium and strain compatibility are simultaneously satisfied; (2) the model can treat wrinkled regions that are non-orthotropic because of asymmetries in the wrinkle's waveform; and (3) the model explicitly provides results for all the elastic constants and the thermal expansion coefficients, of the (monoclinic) wrinkled region. For orthotropic wrinkled regions, a model also is provided for thermal conductivities. The approach is appropriate to wrinkles formed by "cooperative" distortions of reinforcement layers, in which the waveform of each layer is (more or less) in phase with the waveforms of the other layers. Implementation of the analysis is numerical, using a simple specialized finite-element formulation. Examples of the model's predictions are compared to predictions of previous analyses.

KEY WORDS: composite materials, defects, elastic properties, fiber composites, laminates, stress analysis, thermal conductivity, thermal expansion, waves, wrinkles, fatigue (materials), fracture mechanics

Nomenclature[2]

A Amplitude of wrinkle (see Figs. 2 and 8)

C_{jk} Elastic stiffness constant for a slice

$C_{jk}°$ Elastic stiffness constant for the ideal straight material referred to *a-b-c* coordinates (Figs. 2 and 5)

$\overline{C}_{jk}$ Effective elastic stiffness for the wrinkled region (RVE)

---

[1] Research engineer, Jortner Research & Engineering, Inc., Costa Mesa, Calif. 92626.

[2] Notes: *The use of overbars and the ° superscript for these symbols is consistent with the usage for stiffness and strain.

Unless otherwise noted, all quantities are referred to the 1-2-3 coordinate system, Figs. 2 and 5.

Contracted notation is used for stresses and strains, etc., that is, each subscript represents two subscripts in tensor notation: $1 = 11, 2 = 22, 3 = 33, 4 = 23, 5 = 13, 6 = 12$.

E_{jk} Young's modulus in the j-direction*
e_j Strain within a slice, in the j-direction
$\overline{e}_j$ Volume average strain in wrinkled RVE
G_{jk} Shear modulus for jk-shear*
k_{jk} Thermal conductivity of slice, for j-direction heat flow in response to k-direction thermal gradient*
Q_j Heat flow, j-direction*
S_{jk} Elastic compliance constant for slice*
T Temperature rise above the temperature at which RVE slices are free of stress (in absence of external tractions on RVE)
x_j Distance along the j-direction; 1-direction implied if no subscript
α_j Coefficient of thermal expansion of slice, in j-direction*
σ_j Stress, j-direction*
θ Distortion angle, see Figs. 2 and 5
λ Wavelength, or total length of RVE in the 1-direction, see Figs. 2 and 5
ν_{jk} Poisson's ratio, defined as $-S_{jk}E_j$*

Wrinkles sometimes occur in composite materials as a result of improper layup of the reinforcing yarns or layers, or from deformations of the reinforcement during debulking or cure of the matrix. From the stress analyst's viewpoint, wrinkles have been treated, and referred to in the literature, as regions of curved fibers. The thermal conductivities, thermal expansions, and elastic constants of such wrinkled regions can be significantly different from the corresponding properties of a composite with straight reinforcements. Previous work (reviewed in the following section) gives upper or lower bounds to the effects of wrinkling on elastic constants. Unfortunately, the discrepancy between these bounds can be quite large.

This paper provides a new model for estimating the effective thermal and elastic properties of wrinkled regions in fiber-reinforced composite laminates. The wrinkle type for which the modeling approach is appropriate is the relatively common one in which the layers of reinforcement are distorted "cooperatively" so that the waveform of each layer is more or less in phase with the waveforms of neighboring layers. For this type of wrinkle, the model provides numerical estimates of the exact solutions rather than upper or lower bounds. The numerical analysis has been implemented with a FORTRAN-coded computer program called WAVETEC (WAVE Thermal and Elastic Constants). Numerical results are presented for selected cases and compared to predictions of previous models.

Review of Selected Prior Work

Since 1966, attention has been given to predicting the elastic behavior of composites with distorted, or wrinkled, reinforcements. Most of the published results deal in detail only with the composite's Young's modulus, $\overline{E}_1$, in the nominal axial direction of the reinforcement.

Bolotin [1][3] examined the case of a layered material in which the layers are slightly distorted in a random manner (Fig. 1). He arrived, via statistical arguments, at the expression

$$\overline{E}_1 = \frac{E_a}{1 + \left(\dfrac{E_a}{G_{ab}}\right)\phi^2} \tag{1}$$

where ϕ^2 is proportional to the "mean square angle" of the layer distortions; that is, it is a measure of the cumulative effects of the deviation angles, θ, shown in Fig. 1. Bolotin notes that his method may be used to deduce the effects of distortions on the other elastic constants; however, he presents no equations comparable to Eq 1 for the other constants.

Tarnopol'skii et al [2] deal with a regularly distorted layered, or fiber-reinforced, composite. By assuming all the layers, or fibers, conform to a sine wave of uniform amplitude, wavelength, and phase throughout the body (Fig. 2), they are able to provide a deterministic analysis (as opposed to Bolotin's statistical analysis); indeed, they show that the somewhat mysterious ϕ^2 of Eq 1 is

$$\phi^2 = \frac{f^2}{2} \qquad f \equiv \frac{A\pi}{\lambda} \tag{2}$$

for a regularly distorted medium when $(A/\lambda) \ll 1$.

In deriving this result, the composite is treated as a chain of slices of width Δx (Fig. 2). The properties of each slice are the same relative to the local coordinate system (a, b, c) but are different in the body coordinate system (1,

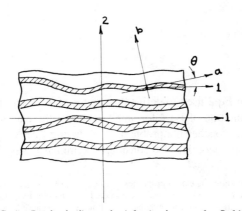

FIG. 1—*Randomly distorted reinforcing layers, after Ref 1.*

[3] The italic numbers in brackets refer to the list of references appended to this paper.

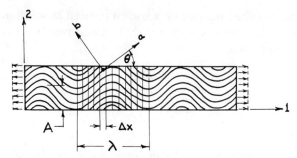

FIG. 2—*Regularly distorted composite, after Ref 2.*

2, 3) because of the rotation through the angle θ.[4] In the 1-direction, the compliance over the entire wavelength (or any integral number of wavelengths) is estimated by assuming that the stress σ_1 would be the same in each slice if the composite body were placed in simple tension or compression along the 1-direction. Then, the total elongation of the composite is simply the sum of the elongations of the slices, and the compliance of the composite is

$$\overline{S}_{11} \equiv \frac{1}{\overline{E}_1} = \frac{1}{\lambda} \int_\lambda \frac{1}{E_1}\, dx \tag{3}$$

The result of the indicated integration for a sinusoidal waviness is

$$\frac{1}{\overline{E}_1} = \frac{1}{E_a}\left(\frac{2 + f^2}{M}\right) + \frac{1}{E_b}\left(1 - \frac{2 + 3f^2}{M}\right) + \left(\frac{1}{G_{ab}} - \frac{2\nu_{ab}}{E_a}\right)\frac{f^2}{M} \tag{4}$$

where

$$f \equiv \frac{A\pi}{\lambda}, \text{ and}$$

$$M \equiv 2(1 + f^2)^{3/2}.$$

Equation 1, with Eq 2 used to define ϕ in terms of f, is said to be a reasonably accurate representation of Eq 4 for small values of the parameter, f.

Nosarev [3] approaches the problem of predicting fiber curvature effects in a somewhat different way. He also considers a symmetric waveform, but does not

---

[4] Note the need to know the elastic properties (E_a, G_{ab},. . .) of a similar composite with absolutely straight reinforcements. All the approaches considered here require that information. In actuality, the properties of this ideal composite may be difficult to obtain because measurements can be made only on real materials, which, as Tarnopol'skii et al point out, may inherently contain distortions of the reinforcement. Thus, one is usually forced to apply micromechanical theory to estimate the properties of the ideal straight composite.

restrict his analysis to sinusoids; instead of closed-form integration, he uses numerical summation. In effect, the waveform is divided into slices (as in Fig. 2), a deviation angle is assigned to each slice, the stiffness matrix of that slice is computed in the body coordinate system (1, 2, 3), and the stiffness of the wavy body is taken as the average of the 1-2-3 stiffnesses of the finite number of slices. Although he only provides numerical results for $\overline{E}_1$, Nosarev notes that all the elastic stiffnesses of the (orthotropic) composite can be found by similar numerical procedures.

Note that Tarnopol'skii et al averaged the compliances, whereas Nosarev averaged the stiffnesses. These two approaches represent lower and upper bounds to the true stiffness of a composite wavy material. The average-compliance method amounts to assuming that stresses are uniform from slice to slice (Fig. 2) when the body is subjected to uniform external tractions. Thus, stress equilibrium is satisfied but there is no guarantee that compatibility of displacements is preserved among the slices. The stiffness calculated thereby is a low bound to the true stiffness, per the Theorem of Minimum Complementary Energy (Ref 4, for example). On the other hand, averaging the stiffnesses amounts to assuming uniform strain throughout the body; compatibility of displacement is satisfied but stress equilibrium might not be obtained. The resulting stiffness is an upper bound estimate of the true stiffness, per the Theorem of Minimum Potential Energy [4].

Both Tarnopol'skii et al and Nosarev provide experimental data to show their respective predictions to be reasonably close to measured stiffnesses. It is instructive to note that the experiments are on different composite materials. The waviness in Tarnopol'skii's composite is the waviness of fiber or fabric layers, closely spaced, so a section through a specimen looks like Fig. 2. In such a case, the composite stiffness in the 1-direction is probably quite close to the lower bound estimated in Ref 2; the composite may be viewed as a stack of slices in which σ_1 must be continuous across the slice boundaries, and the strain, e_1, in each slice may be different without violating compatibility of displacements. Nosarev tested a composite reinforced with somewhat widely spaced wires, each of which was wavy before assembly into the composite, so the phase of each wire's wave is independent of the phase of the other wires' waves (Fig. 3). In this case, each slice (Fig. 3) contains a variety of wire angles; in the limit of a large enough body, each slice can be viewed as containing all possible wire angles, so that the properties of each slice are the same and equal to the average properties of the composite itself. The problem of estimating the properties of each slice is similar to determining the properties of a parallel array of segments of the same straight composite, each at a different angle to the load axis, as sketched in Fig. 4. Clearly, the assumption of the same strain, e_1, in each segment is more appropriate than the assumption of uniform stress. Nosarev's analysis therefore seems appropriate to the material he was studying.

Obviously, it is important to tailor the theoretical assumptions to match the material at hand. As the discrepancy between upper and lower bounds can be

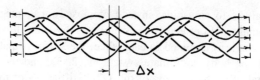

FIG. 3—*Composite reinforced with out-of-phase wavy wires, after Ref 3.*

substantial (see the section on Illustrative Results), we may presume that appli-cation of the Tarnopol'skii model to Nosarev's material, or vice-versa, would not be successful.

Robinson [5] provides an analysis similar to that of Tarnopol'skii et al. Instead of a sinusoidal waveform, Robinson treats a special wave in which, over a quarter wavelength

$$\frac{d\theta}{dx} = Kx \qquad (5)$$

where K is a constant. The full wave is built up of quarter-wave segments bearing the same relationship to each other as the quarter-wave segments of a sinusoid. This waveform has two potential advantages: the integration of the rotated com-pliances can be done in closed form; and the waveform provides an alternate to the sinusoid for representing real distortions in composites. Having defined the waveform, Robinson otherwise follows the compliance-averaging approach adopted by Tarnopol'skii et al, and arrives at

$$\bar{S}_{11} \equiv \frac{1}{E_1} = U + V\frac{\sin 2\theta_m}{2\theta_m} + W\frac{\sin 4\theta_m}{4\theta_m} \qquad (6)$$

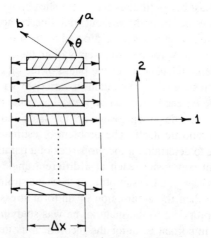

FIG. 4—*Parallel array of segments used to model Nosarev's composite.*

where

$$U \equiv \frac{3}{8}\left(\frac{1}{E_a} + \frac{1}{E_b}\right) + \frac{1}{8}\left(\frac{1}{G_{ab}} - \frac{2v_{ab}}{E_a}\right),$$

$$V \equiv \frac{1}{2}\left(\frac{1}{E_a} - \frac{1}{E_b}\right), \text{ and}$$

$$W \equiv \frac{1}{8}\left(\frac{1}{E_a} + \frac{1}{E_b}\right) - \frac{1}{8}\left(\frac{1}{G_{ab}} - \frac{2v_{ab}}{E_a}\right).$$

and θ_m is the maximum value of θ for the waveform. From the additional formulae Robinson gives in Ref 6, it is easy to derive similar expressions for the other compliance-averaged elastic constants of orthotropic wrinkles that conform to Eq 5.

Bert [7] has extended Tarnopol'skii's method of averaging compliances to provide equations similar to Eq 4 for all the elastic constants of a sinusoid-textured material. His results, in matrix form, are

$$\begin{Bmatrix} \bar{S}_{11} \\ \bar{S}_{12} \\ \bar{S}_{22} \\ \bar{S}_{66} \end{Bmatrix} = \begin{bmatrix} S_{bb} & I_1 & I_2 \\ S_{ab} & I_2 & -I_1 \\ S_{aa} & I; & I_2 \\ \dfrac{1}{G_{ab}} & I_4 & -I_4 \end{bmatrix} \begin{Bmatrix} 1 \\ m^2 \\ \overline{m^4} \end{Bmatrix} \quad (7)$$

where

$$S_{aa} \equiv \frac{1}{E_a} \qquad S_{ab} \equiv \frac{1}{E_b} \qquad S_{ab} \equiv \frac{-v_{ab}}{E_a};$$

$$\begin{Bmatrix} I_1 \\ I_2 \\ I_3 \\ I_4 \end{Bmatrix} \equiv \begin{bmatrix} 0 & 2 & -2 & 1 \\ 1 & -2 & 1 & -1 \\ -2 & 2 & 0 & 1 \\ 4 & -8 & 4 & -4 \end{bmatrix} \begin{Bmatrix} S_{aa} \\ S_{ab} \\ S_{bb} \\ \dfrac{1}{G_{ab}} \end{Bmatrix};$$

$$\overline{m^4} \equiv \frac{1}{\lambda}\int_\lambda \cos^4\theta\, dx = \frac{1 + \dfrac{f^2}{2}}{(1 + f^2)^{3/2}}; \text{ and}$$

$$\overline{m^2} \equiv \frac{1}{\lambda}\int_\lambda \cos^2\theta\, dx = \frac{1}{\sqrt{1 + f^2}}.$$

We may note that Bert's equations provide lower bound estimates of the true stiffnesses of the composite.

In all the approaches reviewed, the analyses deal with symmetric waveforms so that the resulting wavy material is orthotropic; that is, the effective stiffness and compliance matrices of the RVE have no shear-to-extension, or shear-to-shear, coupling terms in the 1-2-3 coordinate system.

A New Numerical Approach

The approach described here predicts thermal expansions, thermal conductivities, and elastic constants for materials of wavy texture. No reference need be made to fibers or layers of cloth. The material analyzed has a repeating volume element (RVE) of the type shown in Fig. 5; its characteristics include: (a) a regular wavy texture; (b) properties independent of translation along the 2- and 3-directions; and (c) properties that vary, with translation along the 1-axis, only because of the rotation of the material locally through the angle, θ; that is, there is postulated an "ideal" straight-textured material that is orthotropic in the a-b-c coordinates, of which the RVE is made. The properties of this ideal material are assumed available, having been measured or calculated theoretically.

As shown in Fig. 5, the wavy RVE is imagined to be divided into a finite number of slices of equal width, Δx. Within the slice, the distortion angle, θ, is assumed constant, and the properties of the slice are those of the ideal material. In the 1-2-3 coordinate system, the properties of the slice may be calculated by rotation of the property matrix through the angle, θ, using well-established equations (such as those of Ref 8). The variation of θ with x is arbitrary; the analysis will be numerical and can deal with (almost) any waveform.

A complication arises if the wavy texture is derived from distortions of fibers or layers that are continuous through the RVE. Simple geometric considerations show that the spacing or thickness of such layers must vary with distortion angle. An implication is that the reinforcement volume fraction will vary with axial (1-

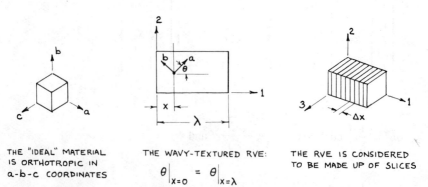

THE "IDEAL" MATERIAL IS ORTHOTROPIC IN a-b-c COORDINATES

THE WAVY-TEXTURED RVE: $\theta\big|_{x=0} = \theta\big|_{x=\lambda}$

THE RVE IS CONSIDERED TO BE MADE UP OF SLICES

FIG. 5—*Idealization of wavy-textured material in which waviness is represented by rotation of an ideal orthotropic material.*

axis) position so that there is no single ideal material whose simple rotations determine the local properties at each axial station. For the time being, this complication is ignored, as it has been by previous investigators, although the numerical method could be extended to include a theoretical approach to defining ideal material properties as a function of distortion angle.

Elastic Constants and Thermal Expansions

The modeling approach is "exact," rather than aiming at a lower or an upper bound, in that stress equilibrium and strain compatibility are both satisfied. However, the results, obtained by a numerical method, are subject to uncertainties associated with the use of a finite slice width Δx. Presumably, as Δx approaches zero, the results converge to the exact solution. To proceed with the analysis, we assume that stress and strain are uniform within each slice but can vary from slice to slice; these stresses and strains must satisfy four requirements: (1) displacement compatibility of neighboring slices, (2) periodicity of displacement at the RVE boundaries (to ensure displacement compatibility between neighboring RVEs), (3) stress equilibrium of each slice, and (4) continuity of stress across contacting slice boundaries; in addition to Hooke's law for the ideal material.

By inspection (see Fig. 6), the following set of constraints satisfies these

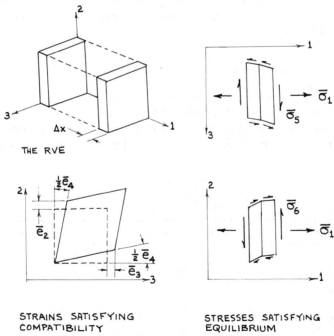

THE RVE

STRAINS SATISFYING
COMPATIBILITY

STRESSES SATISFYING
EQUILIBRIUM

FIG. 6—*Constraints applied to each slice of the wavy-textured RVE.*

conditions

$$\sigma_1 = \bar{\sigma}_1$$

$$e_2 = \bar{e}_2$$

$$e_3 = \bar{e}_3$$

$$e_4 = \bar{e}_4 \tag{8}$$

$$\sigma_5 = \bar{\sigma}_5$$

$$\sigma_6 = \bar{\sigma}_6$$

These constraints define six of the 12 components of stress and strain for the RVE. The remaining six components are defined in the form

$$\bar{e}_k \equiv \frac{1}{\lambda} \sum_\lambda e_k \Delta x; \qquad \bar{\sigma}_k \equiv \frac{1}{\lambda} \sum_\lambda \sigma_k \Delta x \tag{9}$$

Here, the values of e_k or σ_k for each slice i are the response of that slice to any set of constraints (Eq 8) that is applied to the RVE. We also assume the temperature of each slice to be equal to the temperature of the RVE. Generally, for the ith slice

$$\sigma_j = C_{jk}(e_k - \alpha_k T) \tag{10}$$

where the C_{jk} are obtained by rotation of the stiffness matrix $[C^\circ]$ of the ideal material through the texture angle, θ. Equation 10, with double-subscript summation implied, is a set of six simultaneous equations with six unknowns: e_1, e_5, e_6, σ_2, σ_3, and σ_4. Seven special cases of Eq 10 are solved, in sequence, to

TABLE 1—*Seven special problems.*

Problem Number	Nonzero Constraint	Calculated RVE Constants
1	$\bar{\sigma}_1$	$\bar{C}_{11}\ \bar{C}_{16}\ \bar{C}_{66}\ \bar{C}_{26}\ \bar{C}_{12}\ \bar{C}_{13}\ \bar{C}_{36}$
2	$\bar{\sigma}_6$	
3	$\bar{e}_2$	$\bar{C}_{22}\ \bar{C}_{23}$
4	$\bar{e}_3$	$\bar{C}_{33}$
5	$\bar{\sigma}_5$	$\bar{C}_{55}\ \bar{C}_{45}$
6	$\bar{e}_4$	$\bar{C}_{44}$
7	T	$\bar{\alpha}_1\ \bar{\alpha}_2\ \bar{\alpha}_3\ \bar{\alpha}_6$

derive the stiffness matrix and the thermal expansion coefficients of the wavy RVE. These seven cases are described in Table 1.

For example, the first problem solved assumes the application of an arbitrary (unit) uniaxial stress $\bar{\sigma}_1 = 1$ to the RVE; the other constraints of Eq 8 are zero. For this case, Eq 10 is

$$\sigma_1 = C_{11}e_1 + C_{16}e_6 = 1$$

$$\sigma_2 = C_{12}e_1 + C_{26}e_6$$

$$\sigma_3 = C_{13}e_1 + C_{36}e_6 \tag{11}$$

$$\sigma_6 = C_{16}e_1 + C_{66}e_6 = 0$$

$$\sigma_4 = \sigma_5 = e_4 = e_5 = 0$$

Equation 11 can be solved for e_1, e_6, σ_2, and σ_3; thus, after doing the summations of Eq 9, all the stress and strain components for the RVE are known. The nonzero components are $\bar{\sigma}_1$, $\bar{\sigma}_2$, $\bar{\sigma}_3$, $\bar{e}_1$, and $\bar{e}_6$. Then, the constitutive law for the RVE

$$\bar{\sigma}_{jk} = \bar{C}_{jk}(\bar{e}_k - \bar{\alpha}_k T) \tag{12}$$

may be expressed in matrix form (for this case) as

$$\begin{Bmatrix} 1 \\ \bar{\sigma}_2 \\ \bar{\sigma}_3 \\ 0 \end{Bmatrix} = \begin{bmatrix} \bar{C}_{11} & \bar{C}_{16} \\ \bar{C}_{12} & \bar{C}_{26} \\ \bar{C}_{13} & \bar{C}_{36} \\ \bar{C}_{16} & \bar{C}_{66} \end{bmatrix} \begin{Bmatrix} \bar{e}_1 \\ \bar{e}_6 \end{Bmatrix} \tag{13}$$

This set of four equations is insufficient to solve for the seven unknown terms of $\bar{C}_{jk}$. The additional necessary equations may be provided by going on to the second problem, in which all the constraints of Eq 8 are zero except $\bar{\sigma}_6 = 1$. The same procedure as in the first case yields

$$\begin{Bmatrix} 0 \\ \bar{\sigma}'_2 \\ \bar{\sigma}'_3 \\ 1 \end{Bmatrix} = \begin{bmatrix} \bar{C}_{11} & \bar{C}_{16} \\ \bar{C}_{12} & \bar{C}_{26} \\ \bar{C}_{13} & \bar{C}_{36} \\ \bar{C}_{16} & \bar{C}_{66} \end{bmatrix} \begin{Bmatrix} \bar{e}'_1 \\ \bar{e}'_6 \end{Bmatrix} \tag{14}$$

where the stress and strain terms are distinguished by the prime (') from the numbers obtained in Case 1 for the same components.

The eight simultaneous equations of Eq 13 and Eq 14 may be solved to give

$$\overline{C}_{11} = \left(p_1 - \frac{q_1 p_6}{q_6}\right)^{-1} \qquad \overline{C}_{12} = (p_2 - \overline{C}_{26} p_6)\frac{1}{p_1}$$

$$\overline{C}_{16} = -\overline{C}_{11}\frac{q_1}{q_6} \qquad \overline{C}_{36} = \frac{q_3 - p_3 r}{q_6 - p_6 r}$$

$$\overline{C}_{66} = (1 - \overline{C}_{16} q_1)\frac{1}{q_6} \qquad \overline{C}_{13} = (p_3 - \overline{C}_{36} p_6)\frac{1}{p_1} \tag{15}$$

$$\overline{C}_{26} = \frac{q_2 - p_2 r}{q_6 - p_6 r}$$

where, for simplicity in notation,

$$p_1 \equiv \overline{e}_1 \qquad q_1 \equiv \overline{e}'_1 \qquad r \equiv \frac{q_1}{p_1}$$

$$p_2 \equiv \overline{\sigma}_2 \qquad q_2 \equiv \overline{\sigma}'_2$$

$$p_3 \equiv \overline{\sigma}_3 \qquad q_3 \equiv \overline{\sigma}'_3$$

$$p_6 \equiv \overline{e}_6 \qquad q_6 \equiv \overline{e}'_6$$

Problem 3 involves setting all constraints to zero except $\overline{e}_2 = 1$. After solving Eq 10 for each slice and doing the summations of Eq 9, the nonzero stress and strain components of the RVE are found ($\overline{e}_1$, $\overline{e}_2 = 1$, $\overline{\sigma}_2$, $\overline{\sigma}_3$, and $\overline{e}_6$), and application of Eq 12 gives

$$\overline{C}_{22} = \overline{\sigma}_2 - \overline{C}_{12}\overline{e}_1 - \overline{C}_{26}\overline{e}_6$$

$$\overline{C}_{23} = \overline{\sigma}_3 - \overline{C}_{13}\overline{e}_1 - \overline{C}_{36}\overline{e}_6 \tag{16}$$

where $\overline{C}_{12}$, $\overline{C}_{26}$, $\overline{C}_{13}$, and $\overline{C}_{36}$ have been determined previously in Eq 15.

Problem 4, in which $\overline{e}_3 = 1$ is the only nonzero constraint, is similarly solved to give

$$\overline{C}_{33} = \overline{\sigma}_3 - \overline{C}_{13}\overline{e}_1 - \overline{C}_{36}\overline{e}_6 \tag{17}$$

Likewise, Problem 5 ($\overline{\sigma}_5 = 1$) gives

$$\overline{C}_{55} = \frac{1}{\overline{e}_5}$$

(18)

$$\overline{C}_{45} = \frac{\overline{\sigma}_4}{\overline{e}_5}$$

and Problem 6 $(\overline{e}_4 = 1)$ gives

$$\overline{C}_{44} = \overline{\sigma}_4 - \overline{C}_{45}\overline{e}_5$$

(19)

The complete (monoclinic) $\overline{C}$ matrix is now known.

The last problem, to derive the effective coefficients of thermal expansion for the RVE, involves setting $T = 1$ and setting to zero all the constraints of Eq 8. The solution to Eq 10 provides the nonzero stress and (total) strain components of each slice, which are then summed (Eq 9) to give $\overline{e}_1$, $\overline{e}_5$, $\overline{e}_6$, $\overline{\sigma}_2$, $\overline{\sigma}_3$, and $\overline{\sigma}_4$. It is then straightforward, using Eq 12, to obtain

$$[\overline{\alpha}] \equiv \begin{Bmatrix} \overline{\alpha}_1 \\ \overline{\alpha}_2 \\ \overline{\alpha}_3 \\ 0 \\ 0 \\ \overline{\alpha}_6 \end{Bmatrix} = [\overline{C}]^{-1}[\overline{\sigma} - \overline{Ce}]$$

(20)

Thermal Conductivities of Orthotropic RVEs

A numerical approach to the exact solution for the thermal conductivities of the wavy RVE could be patterned on the approach just described for the elastic constants and thermal expansions. Here, we present a solution restricted to orthotropic RVEs; that is, the waveform of the texture must be symmetric so that, in the 1-2-3 coordinate system, the RVE conductivity is characterized fully by $\overline{k}_{11}$, $\overline{k}_{22}$, and $\overline{k}_{33}$.

The slices, however, are characterized by k_{11}, k_{22}, k_{33}, and k_{12}, obtained by rotation of conductivity tensor of the ideal material

$$k_{11} = k_{aa} \cos^2 \theta + k_{bb} \sin^2 \theta$$

$$k_{22} = k_{aa} \sin^2 \theta + k_{bb} \cos^2 \theta$$

(21)

$$k_{33} = k_{cc}$$

$$k_{12} = k_{aa} \cos \theta \sin \theta - k_{bb} \cos \theta \sin \theta$$

We assume that the temperature gradient is uniform within each slice, as is the heat flow. To ensure continuity of temperature and heat flow among slices, it is sufficient (by inspection) to require

$$Q_1 = \overline{Q}_1$$

$$\frac{\partial T}{\partial x_2} = \frac{\overline{\partial T}}{\partial x_2} \tag{22}$$

$$\frac{\partial T}{\partial x_3} = \frac{\overline{\partial T}}{\partial x_3}$$

The assumed symmetry of the waveform causes the net effect of cross-coupling via k_{12} to be zero. Thus, with due consideration for the basic definition of thermal conductivity, we can readily derive the following simple expressions for the thermal conductivities of the RVE

$$\overline{k}_{11} = \left(\frac{1}{\lambda} \sum_\lambda \frac{\Delta x}{k_{11}} \right)^{-1}$$

$$\overline{k}_{22} = \frac{1}{\lambda} \sum_\lambda k_{22} \Delta x \tag{23}$$

$$\overline{k}_{33} = \frac{1}{\lambda} \sum_\lambda k_{33} \Delta x$$

Illustrative Results

To demonstrate the nature of the results obtainable with the approach presented, several cases are analyzed and the results compared to those obtained by previous (bounding) approaches [2,3,5,7].

First, the current approach was applied to some of the cases for which closed-form solutions have been given by Robinson in Ref 5. Robinson's cases represent a baseline material (properties listed in Table 2) and three variations designed

TABLE 2—*Hypothetical properties of ideal materials (arbitrary units).*

Case Name	E_a	E_b	E_c	v_{bc}	v_{ac}	v_{ab}	G_{bc}	G_{ac}	G_{ab}
Baseline	12.0	2.0	12.0	0.0	0.0	0.1	1.0	1.0	1.0
High G_{ab}	12.0	2.0	12.0	0.0	0.0	0.1	1.0	1.0	2.0
Low G_{ab}	12.0	2.0	12.0	0.0	0.0	0.1	1.0	1.0	0.1
High E_b	12.0	12.0	12.0	0.0	0.0	0.1	1.0	1.0	1.0

to explore the effects of changes in shear modulus, G_{ab}, and transverse modulus, E_b, on wavy RVE properties. The wrinkle waveforms in these cases conform to Eq 5. Robinson's closed-form "low-bound" results for RVE axial modulus, $\bar{E}_1$, as a function of wrinkle severity are shown in Fig. 7. Also shown in Fig. 7 are the results of our numerical "exact" calculations for the same ideal material and wrinkle geometries. Sixty slices were used to represent the RVE for the numerical analyses; subsequent exploratory calculations showed that about 20 slices are sufficient in similar cases to give a solution that does not change upon increasing the number of slices. As expected, the numerical "exact" predictions of axial stiffness equal or exceed in magnitude the closed-form low-bound predictions.

It appears in Fig. 7 that when the shear modulus, G_{ab}, is not "too low," and when the transverse modulus, E_b, is relatively low, as in the baseline case, the low-bound and exact solutions for $\bar{E}_1$ are in excellent agreement. However, the low-bound solution may be significantly in error when the shear modulus is reduced or the transverse stiffness increased.

The second set of examples is for sinusoidal wrinkles in the same ideal material (low-shear-modulus case in Table 2). The numerical "exact" results are compared to the lower bound and upper bound predictions derived from uniform stress and uniform strain assumptions. The lower bound predictions are essentially the same as Bert's (Eq 7) although the implementation was by numerical

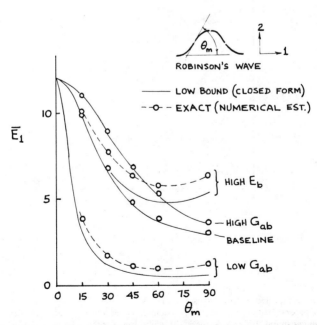

FIG. 7—*Comparison of current numerical-exact solutions to Robinson's closed-form low-bound solutions, for the effective axial Young's moduli of wrinkles in the materials listed in Table 2.*

integration rather than closed-form calculation. The upper bound analysis is similar to Nosarev's (Ref *3*), extended to all the elastic constants. Some results are plotted in Figs. 8 through 11, as a function of sinusoid amplitude/wavelength ratio. The points (circles) in these graphs are the numerical estimates; the curves, which were drawn to help one follow the trends, do not necessarily give accurate interpolations between points.

Note, in Figs. 8 and 9, that the exact solutions for the Young's moduli are rather closer to the lower bound predictions than to the upper bounds.[5] However, neither upper nor lower bound analyses do a good job in predicting the wavy composite's shear modulus (Fig. 10).

Trends for the effective thermal properties of the low-shear-modulus material are given in Fig. 11, assuming $\alpha_a = \alpha_c = 1.0$, $\alpha_b = 2.0$, $k_{aa} = k_{cc} = 1.0$, and $k_{bb} = 0.5$.

Applicability and Limitations

The model presented here provides numerical estimates of the exact solutions for the effective thermal and elastic constants of regions of wrinkled texture in

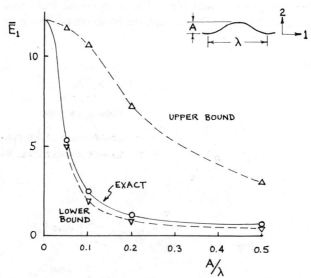

FIG. 8—*Effective axial Young's moduli of sinusoidally wrinkled regions in the low-G_{ab} composite of Table 2.*

---

[5] It is to be expected that $\bar{E}_1$ is closer to the lower bound (see discussion of Tarnopol'skii's and Nosarev's papers). However, that $\bar{E}_2$ also is closer to the lower bound was a surprise because the 2-direction constraint (Eq 8) is $\bar{e}_2$ *not* $\bar{\sigma}_2$. This result may perhaps be viewed as a demonstration of the importance of shear coupling and the fact that the model (properly) allows each slice to have an independent shear strain e_6, whereas the upper bound calculation constrains e_6 to be the same in each slice.

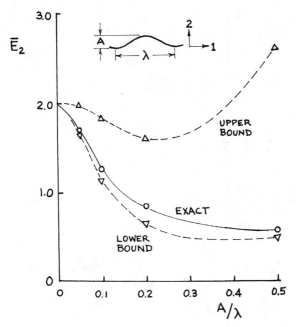

FIG. 9—*Effective transverse Young's moduli of sinusoidally wrinkled regions in the low-G_{ab} composite of Table 2.*

an otherwise orthotropic material. As such, it may be considered an improvement on the prior work [1,2,3,5,7]. However, there remain significant limitations to its applicability.

First, as already pointed out, the model is appropriate to cooperatively wrinkled laminates (Fig. 2) and not appropriate to out-of-phase wrinkling (Fig. 3). That is, the waviness of reinforcing yarns in woven-cloth laminates is not well treated. Such interwoven reinforcements are perhaps closer to Nosarev's model and might be modeled better with an upper-bound (stiffness-averaging) analysis. Also, special efforts may be necessary to deal with the bidirectional waviness in cloth.

Second, the derivation of the constraints (Eq 8) depends on the idea that the RVE is indeed a repeating element in a much larger body. Thus, the model is applicable strictly only to extended three-dimensional solids that are of the same wrinkled texture throughout. Thin wall components are not treated exactly; neither are bodies containing only local areas, or channels, of wrinkled material.

Third, wrinkles in unbalanced laminates also are not treated by the present model because no consideration is given to bending moments. Likewise, unbalance that might be introduced by a wrinkle in a thin section of an otherwise balanced laminate also is ignored.

Thus, many cases of practical interest are not rigorously treated by the proposed model, although the results may be useful as approximations. In judging the severity of these limitations, it may be appropriate to recall that none of the other simple models, described in the literature just reviewed, escape such lim-

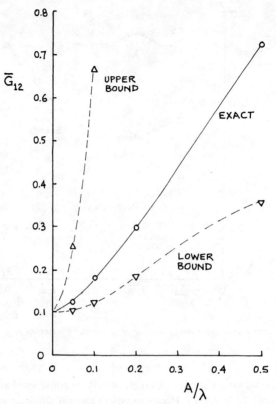

FIG. 10—*Effective shear moduli ($\overline{G}_{12}$) of sinusoidally wrinkled regions in the low-G_{ab} composite of Table 2.*

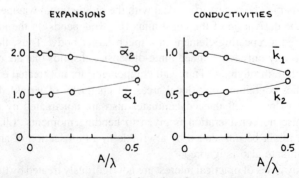

FIG. 11—*Effective thermal expansions* (left) *and conductivities* (right) *of wrinkled regions in the low-G_{ab} composite of Table 2.*

itations. Indeed, the second limitation is inherent in the definition of effective moduli, even with reference to unwrinkled laminates [9].

To predict the behavior of wrinkles in thin sections, modeling may be improved (perhaps) by modifying the constraints of Eq 8 to provide zero stress in the 2-direction ($\sigma_2 = \bar{\sigma}_2 = 0$, rather than $e_2 = \bar{e}_2$), assuming that the 2-direction is the through-thickness direction.

To apply the model to the structural analysis of a wrinkled component, using a finite element method, the element mesh would be defined and material properties assigned to each element. These properties, for elements containing wrinkles, could be estimated from the properties of unwrinkled regions, using the proposed numerical analysis. To do this, the wrinkled regions first would be characterized geometrically by assigning a distortion angle to each location (imaginary slice) within each element of the mesh. Implementing the model would provide the effective properties of the entire element. The level of detail provided by the model would seem most appropriate to use with linear elements; structural analyses with quadratic or cubic elements that allow property and strain variations within an element might benefit from more detailed property models—especially if the elements' dimensions are large relative to the stress/strain gradients in the structure. A potentially significant advantage of the current simple model over previous simple models is the ability to produce a monoclinic property matrix for unsymmetric wrinkles. This allows the prediction of individual properties for finite elements (or regions within a higher-order element), which may be of small dimensions relative to the wrinkle wavelength, as would be appropriate for situations of appreciable strain gradient over a single wrinkle.

Summary

Previous efforts to predict the effective elastic behavior of curved-fiber composites, from the properties of an ideal (orthotropic) straight-fiber composite, rely on two assumptions (among others). First, it is assumed that the stresses, or the strains, are uniform throughout the representative repeating-volume-element (RVE) of the wrinkled material; therefore either stress equilibrium or strain compatibility is satisfied (but not both) so the predictions represent lower or upper bounds, respectively, to the elastic stiffnesses, rather than direct estimates. Second, the waveform of the wrinkle is assumed symmetric about the half-wavelength plane to ensure that the RVE is orthotropic; real wrinkles do not necessarily conform to this assumption.

This paper describes a new theoretical model (implemented numerically) that extends the previous work in several ways: the approach is "exact" for a certain class of wrinkles, within the uncertainty attributable to the numerical implementation, because stress equilibrium and strain compatibility are simultaneously satisfied; the model can treat wrinkled regions that are non-orthotropic (monoclinic) as a result of asymmetries in the wrinkle's waveform; and the model explicitly provides estimates of all the elastic constants and the thermal expansion coefficients of the wrinkled composite. The approach is appropriate for wrinkles in which the layers of reinforcement are distorted cooperatively, so the waveform of each layer is more-or-less in phase with the waveforms of the other layers. For orthotropic wrinkles, a formulation for predicting the thermal conductivities also is described.

Numerical results show that the discrepancies can be large between upper bound stiffness predictions (via the assumption of uniform microstrain) and lower bound predictions (uniform microstress). The estimates of exact behavior provided by the new model can differ substantially from either bound.

The model applies rigorously only to a restricted set of circumstances. It would be desirable to know how well it approximates situations of technical interest to which it does not apply rigorously. Comparisons with experimental data, or with results of detailed finite-element models of wrinkles, are recommended to explore its wider applicability.

Acknowledgments

A major portion of the work was performed while the author was at Science Applications, Inc., Irvine, Calif. The sponsorship of the Office of Naval Research and of the Air Force Wright Aeronautical Laboratories (Materials Laboratory) is gratefully acknowledged. The author thanks W. C. Loomis, N. J. Pagano, E. Y. Robinson, and E. L. Stanton for the various interesting and helpful discussions, and M. J. White for help with computer programming.

References

[1] Bolotin, V. V., *Polymer Mechanics (Mekhanika Polimerov)*, Vol. 2, No. 1, 1966, pp. 7–11 (11–19).
[2] Tarnopol'skii, Yu. M., Portnov, G. G., and Zhigun, I. G., *Polymer Mechanics (Mekhanika Polimerov)*, Vol. 3, No. 2, 1967, pp. 161–166 (243–249).
[3] Nosarev, A. V., *Polymer Mechanics (Mekhanika Polimerov)*, Vol. 3, No. 5, 1967, pp. 567–570 (858–863).
[4] McClintock, F. A., "Approximate Stress Analysis," *Mechanical Behavior of Materials*, F. A. McClintock and A. S. Argon, Eds., Addison-Wesley, Chapter 10, 1966.
[5] Robinson, E. Y., "Approximate Analysis of Wrinkle Anomalies in Carbon-Carbon Material," Technical Memo ATM-78-3451-10-19, Aerospace Corporation, El Segundo, Calif., 24 Jan. 1978.
[6] Robinson, E. Y. and Jortner, J., "Analysis of Wrinkles in Carbon-Carbon Composites," paper presented to 3rd Joint Army-Navy-NASA-Air Force (JANNAF) Rocket Technology Subcommittee Symposium, Langley Field, Va., Oct. 1981.
[7] Bert, C. W. in *Mechanics of Bimodulus Materials*, C. W. Bert, Ed., Publication AMD-Vol. 33, American Society of Mechanical Engineers, 1979, pp. 17–28.
[8] Hearmon, R. F. S., *An Introduction to Applied Anisotropic Elasticity*, Oxford University Press, 1961, Chapter 1.5.
[9] Pagano, N. J. in *Mechanics of Composite Materials*, Vol. 2, G. P. Sendeckyj, Ed., Academic Press, New York, 1974, Chapter 1.

Victoria Papaspyropoulos,[1] *Jalees Ahmad,*[1] *and Melvin F. Kanninen*[1]

A Micromechanical Fracture Mechanics Analysis of a Fiber Composite Laminate Containing a Defect

REFERENCE: Papaspyropoulos, V., Ahmad, J., and Kanninen, M. F., "A Micromechanical Fracture Mechanics Analysis of a Fiber Composite Laminate Containing a Defect," *Effects of Defects in Composite Materials, ASTM STP 836,* American Society for Testing and Materials, 1984, pp. 237–249.

ABSTRACT: The vast majority of developments in fracture mechanics have been instigated by the need to avoid fracture of metals. Consequently, these developments are based on the assumption that the material is homogeneous—an assumption that is far more palatable in the case of most metals than it is for fiber-reinforced composites. Attempts to make fracture mechanics concepts more applicable to composite materials has resulted in the emergence of anisotropic fracture mechanics approaches. But, while this simple extension of linear elastic fracture mechanics does remove the assumption of material isotropy, it still assumes the material to be homogeneous. Experimental evidence suggests that the application of anisotropic fracture mechanics to fiber-reinforced composites still leads to inconsistencies in the prediction of fracture.

This paper presents an improvement in the application of a concept that allows for direct consideration of heterogeneous material behavior in the near crack tip region. The concept is similar to the singular perturbation and matched asymptotic expansion techniques used in fluid mechanics. The problem of a fiber-reinforced composite containing a flaw is divided into a local heterogeneous region (LHR) and a global anisotropic homogeneous continuum region. In the LHR, local failure events can be individually modeled, and then related to a global fracture parameter. Specifically, a quasi-three-dimensional finite-element LHR model is used to investigate micromechanical failure events at the tip of a crack in a fiber-reinforced composite laminate. These events are then related to a global fracture event, whereupon the fracture toughness of the composite can be estimated. Computational results, although in seemingly reasonable agreement with existing experimental data, are heuristic in nature. The work represents only a preliminary effort toward possible future development of a more sophisticated predictive model.

KEY WORDS: fracture mechanics, fiber-reinforced composites, finite-element method, fatigue (materials), composite materials

In the absence of a significant stress riser like a sharp crack, small inherent defects in a fiber composite material (for example, broken fibers, matrix flaws,

[1]Research engineer, senior research engineer, and institute scientist, Engineering and Material Sciences Division, Southwest Research Institute, San Antonio, Tex. 78284.

and debonded interfaces) can coalesce to form an identifiable crack-like flaw. As it approaches a critical size, the size scale of the damage zone near the crack tip will still be small relative to the crack length and other dimensions of the body. Hence, micromechanical failure mechanisms, such as fiber breakage, debonding, and matrix cracking, control the fracture process. This will be true whether the composite is composed of uniaxially or multiaxially oriented fibers. Figure 1 illustrates the many local failure events that can contribute to the damage growth that generally precedes fracture in a fiber composite.

Kanninen et al [1][2] have taken the point of view that a proper fracture mechanics analysis must treat the micromechanical damage mechanism directly. To achieve this end, they developed a model that can be linked conceptually to the singular perturbation and matched asymptotic expansion techniques used in fluid mechanics. That is, the problem of a composite material containing a flaw is divided into a local and global region. In the local region surrounding the crack tip, called the Local Heterogeneous Region (LHR), each constituent of the composite is modeled separately and given its own individual mechanical and fracture properties. Surrounding this region is the global region. This is modeled as an anisotropic homogeneous continuum and represents the bulk

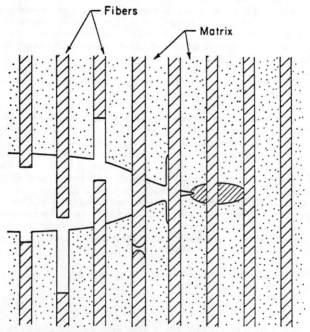

FIG. 1—*Model of a crack tip in a fiber composite illustrating the various energy dissipation mechanisms involved in crack propagation.*

[2]The italic numbers in brackets refer to the list of references appended to this paper.

response characteristics of a single lamina or a unidirectional laminate. A load (or displacement) applied at a location remote from the LHR is transmitted through the global region to the boundary of the LHR. Failure mechanisms inside the LHR are then influenced by the applied stress, the component geometry, and the local stress concentration (due to the macroscopic crack length), as well as by the deformation and failure of neighboring constituents.

A typical LHR for a unidirectional fiber composite containing three distinct components (that is, the fibers, the matrix, and the fiber matrix interface zones) is shown in Fig. 2. In the approach used in Ref *1*, any of these components ruptures when an intrinsic critical energy dissipation rate is reached. The far-field solution was made to reflect damage increases in the LHR by increasing the crack length. The global model then provides a new set of boundary conditions for the LHR.

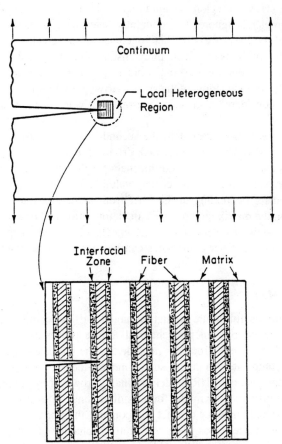

FIG. 2—*The analysis model of Kanninen et al [1] with a local heterogeneous region embedded in a linear elastic anisotropic continuum.*

The heuristic computations presented in Ref *1* were performed using a two-dimensional idealization of the problem. The finite-element analysis used constant strain triangular elements. This paper represents a two-fold improvement over that stage of the model. First, a three-dimensional LHR model containing an ordered array of fibers is constructed. Second, a more accurate isoparametric finite-element formulation is used. However, because of unavailability of sufficient material property data, the computational results are still largely heuristic in nature.

Model Development

Composite fracture research results given in the literature generally fall into one of two broad categories. These are either (1) a continuum analysis for a homogeneous anisotropic linear elastic material containing an internal or external flaw of known length, or (2) a semiempirical analysis of the micromechanical details of the crack-tip region in a unidirectional fiber composite. The continuum approach completely ignores the inherently heterogeneous nature of composite materials and the basic way that heterogeneity affects crack extension. In fact, this approach represents only a slight extension of ordinary linear elastic fracture mechanics to account for the anistropic response of the material to load. It involves only an evaluation of the crack driving force with its critical value tacitly being considered a material constant that can be obtained from experiments.

The micromechanical approach, the second of the two just cited, essentially represents a way to determine the crack driving force in terms of basic material properties by considering the various mechanisms involved in composite fracture. For example, values of G_c, the critical value of the energy release rate, have been deduced for debonding of the fiber from the matrix material, pull-out of the fiber from the matrix, and for inelastic deformation and fracture of the matrix material. A review of several micromechanical and continuum approaches to the failure prediction of fiber composites can be found in a paper by Kanninen et al [2].

Basis of the Model

A practically useful failure prediction model for fiber composites must reflect the role of the various micromechanical failure processes. Furthermore, the combined effect of these processes must dictate the ultimate macroscopic failure point of the composite. To devise such a model, the present development relies on the two-dimensional LHR model originally proposed by Kanninen et al [1]. A conceptual view of their model for a unidirectional 0-deg ply is shown in Fig. 2. The model contains three distinct components: the fibers, the matrix, and the fiber-matrix interface zones. The constitutive behavior of each of these components must be prescribed. In addition, failure criteria, such as an intrinsic critical energy dissipation rate, for each component must be known.

The LHR is modeled over a small area surrounding the crack tip. The failure events inside the LHR are assumed to be controlled by a far field fracture mechanics parameter; for example, by the stress intensity factor, if the region surrounding the LHR behaves in a linear elastic fashion. This outer region is viewed as a homogeneous anisotropic continuum in which the simple "rule of mixtures" relationships for defining the elastic moduli can be used.

The computation begins by finding the boundary condition for the LHR. This is done by solving the global problem in which the material is taken to be homogenous. These boundary conditions are then prescribed in the local analysis involving the LHR. The size of the LHR is chosen to be large enough so that the local failure events do not significantly change the conditions at its boundary, but yet sufficiently small to allow accurate modeling of each component without excessive computational expense.

In Ref 1, computations were performed for arbitrary flaw size and orientation for unidirectional composites with linear elastic-brittle constituent behavior. The mechanical properties were nominally those of graphite epoxy. With the rupture properties arbitrarily varied to test the capability of the model to reflect real fracture modes in fiber composites, it was shown that fiber breakage, matrix crazing, crack bridging, matrix-fiber debonding, and axial splitting can all occur during a period of (gradually) increasing load prior to catastrophic fracture. Qualitative comparisons with experimental results of Brinson and Yeow [3] on edge-notched unidirectional graphite/epoxy specimens were also made. In Ref 1, computations were performed using a two-dimensional finite-element code. Therefore, each element of the LHR was assumed to be continuous in the out-of-plane direction (Fig. 3a). This, of course, is unrealistic. In actual composites, the fibers are randomly distributed through the thickness (Fig. 3b).

If the number of fibers is not very large, one could perhaps use a three-dimensional finite-element code to model each individual fiber without excessive computer storage (and cost) requirements. This not being the case, a compromise between precise modeling and computational effort is necessary. A reasonable compromise is to consider an ordered array of elements containing fibers (Fig. 3c). Then, a reduction in computational effort is possible if one can ignore the free surface effects at $Z = \pm t$ (Fig. 4a). For this case, only one layer of thickness, Δt, removed from the thickness direction (Fig. 4b) needs to be considered with appropriate boundary conditions to enforce periodicity in the Z-direction.

The geometry of this repeating layer also suggests that, in a finite-element model, it can be represented by three types of special elements, as shown in Fig. 5a. If the stiffness of the interface can be absorbed into the stiffness of the fiber, then only one formulation of a sandwich-type element is needed. This is shown in Fig. 5b. Type 3 element is simply a special case when Material 1 is the same as Material 2.

Another essential feature of the micromechanical approach is the modeling of

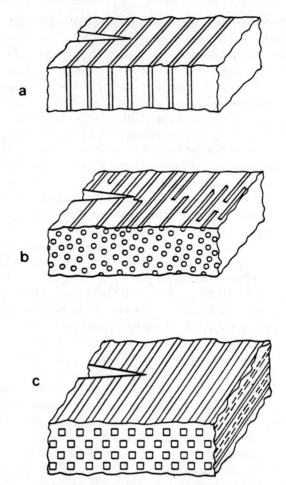

FIG. 3—*Representations of fiber distribution in three dimensions:* (a) *two-dimensional,* (b) *random, and* (c) *idealized three-dimensional.*

the failure mechanisms. In their two-dimensional model Kanninen et al [*1*] used spring-like connections between node points that were severed once a prescribed local failure criterion was met. In the present three-dimensional case, this technique would lead to excessive computer storage requirements, and possibly to an ill-conditioned global stiffness matrix. To circumvent this situation, a method successfully used by Bazant [*4*] for reinforced concrete structures may be employed. In this method, each element in which a prescribed failure criterion is satisfied is made to lose its stiffness. This loss of stiffness or ''death'' of a particular element is enforced by simply deleting the corresponding terms in the global stiffness matrix. The element death option is particularly appealing when used in conjunction with the special sandwich element discussed earlier.

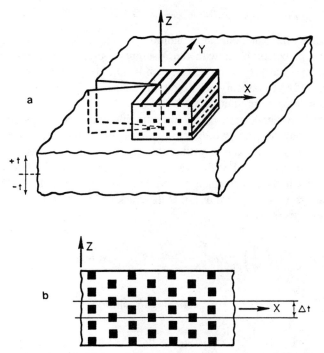

FIG. 4—(a) *Three-dimensional view of the local heterogeneous region and* (b) *a repeating layer.*

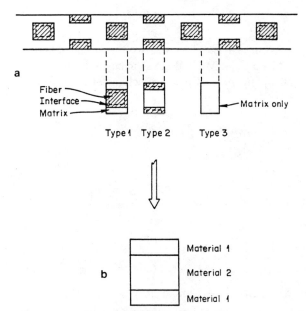

FIG. 5—*Conceptual development of special elements for fiber-reinforced composites:* (a) *three-element types, and* (b) *sandwich element.*

Determination of Boundary Conditions

As stated earlier, the boundary conditions for the LHR are prescribed by first solving the global problem with the assumption of homogeneous anisotropic equivalent material properties. Consequently, the displacement boundary conditions at the LHR boundary are exactly the same as if the entire cracked body were an anisotropic continuum.

For many problems of practical interest, this phase of computation can be performed using simpler two-dimensional analysis procedures. For example, consider an elastic continuum with rectilinear anisotropic properties subjected to in-plane deformation. For this case, the work of Sih and Liebowitz [5] can be used to obtain the displacement boundary conditions for the LHR. For a polar coordinate system (r, θ) with origin at the crack tip, the displacements near the crack tip are given by

$$
\begin{aligned}
u_i = K_{\mathrm{I}} \left(\frac{2r}{\pi} \right)^{1/2} Re &\left\{ \frac{1}{S_1 - S_2} [S_1 p_{2i} (\cos \theta + S_2 \sin \theta)^{1/2} \right. \\
&\left. - S_2 p_{2i-1} (\cos \theta + S_1 \sin \theta)^{1/2}] \right\} \\
+ K_{\mathrm{II}} \left(\frac{2r}{\pi} \right)^{1/2} Re &\left\{ \frac{1}{S_1 - S_1} [p_{2i} (\cos \theta + S_2 \sin \theta)^{1/2} - p_{2i-1} (\cos \theta \right. \\
&\left. + S_1 \sin \theta)^{1/2}] \right\}
\end{aligned}
\tag{1}
$$

where i takes the values 1 and 2 for the two displacement components, and K_{I} and K_{II} are the Mode I and Mode II stress intensity factors.

The constants p_1, p_2, p_3, and p_4 are functions of the elastic constants given by

$$
\begin{aligned}
p_1 &= a_{11} S_1^2 + a_{12} - a_{16} S_1 \\
P_2 &= a_{11} S_2^2 + a_{12} - a_{16} S_2 \\
p_3 &= \frac{1}{S_1} (a_{12} S_1^2 + a_{22} - a_{26} S_1) \\
p_4 &= \frac{1}{S_2} (a_{12} S_2^2 + a_{22} - a_{26} S_2)
\end{aligned}
\tag{2}
$$

where S_1, S_2, $\bar{S}_1$, and $\bar{S}_2$ are the roots of the characteristic equation

$$
a_{11} S^4 - 2a_{16} S^3 + (2a_{12} + a_{66}) S^2 - 2a_{26} S + a_{22} = 0
\tag{3}
$$

In the special case when the material orthotropy Directions 1 and 2 are parallel and normal to the crack line, the elastic constants, a_{1k}, can be given as

$$a_{11} = \frac{1}{E_1}, \; a_{22} = \frac{1}{E_2}, \; a_{66} = \frac{1}{\mu_{12}}$$

$$a_{12} = -\frac{\nu_{12}}{E_1} = -\frac{\nu_{21}}{E_2} \tag{4}$$

$$a_{16} = a_{26} = 0$$

where E, μ, and ν stand for extensional modulus, shear modulus, and Poisson's ratio, respectively.

The case discussed in the preceding discussion is only a simple example. The use of the LHR concept is, of course, not restricted to this simple situation. In general, the LHR boundary conditions can be found by performing a three-dimensional finite-element stress analysis using a homogeneous anisotropic equivalent material description. Further, in the development presented so far, it has been assumed that the boundary conditions for the LHR remain essentially unchanged. Hence, the local damage processes of its constituents can be modeled only as long as the damage zone remains small compared to the LHR size. This assumption, while appropriate for the heuristic nature of the present study, is not essential to the LHR concept. It is used simply as an expedient to reduce the computational effort. In general, sequential updating of the boundary conditions following each local damage event can be used.

Constituent Material Properties

The material properties of each constituent in the LHR, as well as the equivalent material properties for the global analysis presented in this paper, were assumed to be linear elastic. Once again, this assumption is not essential to the model. It is used here so that the model can be exercised without excessive computational expense. In general, nonlinear and rate-dependent material properties of the constituents may be treated. In that case, however, the use of linear elastic fracture mechanics concepts will be inappropriate, adding further complications to the computational process.

In keeping with the primary purpose of the LHR model development, each of the individual constituents of the composite must be capable of rupturing to allow the body to exhibit the changes in strength that correspond to various levels and orientations of local damage. In the present stage of model development, this is done by using the element death option in the finite-element analysis. A similar technique has been effectively used by Bazant [4] in the study of fracture in reinforced concrete structures. In general, the technique may be employed to selectively reduce or eliminate the stiffness of a given finite element in any given direction when a local failure criterion is met. In the present work, however, it is assumed (for simplicity) that once the average strain energy of a given element exceeds a prescribed critical value, the element loses all components of its stiffness simultaneously. Since each member of the LHR, for example, each fiber, is made up of several finite elements, this element death implies a very local loss of stiffness.

In the present work, as in Ref *1*, the material properties used are those for graphite fibers and epoxy matrix deduced from the experimental work of Brinson and Yeow [*3*]. These properties are contained in Table 1. The properties of the interface between fibers and epoxy could not be directly obtained from Ref *3*. Therefore, in the computations, they were simply assumed to be the same as those of the matrix. In essence, this means that the LHR is composed only of matrix and fibers.

For the global analysis, the following equivalent material properties were used: E_1 = 124105 MPa, E_2 = 4757 MPa, v_{12} = 0.3, and μ_{12} = 20684 MPa, where Direction 1 corresponds to the fiber direction. In the numerical computations performed in this work two fiber orientations were considered: (*a*) the fibers were assumed to be normal to the crack line, and (*b*) the fibers were assumed to be parallel to the crack line. For each case, the global analysis was performed using Eq 1 with appropriate material properties to determine the LHR boundary conditions.

Example Computational Results

To test the validity of the three-dimensional LHR model concept, computations were performed assuming the crack tip as belonging to a double-edge-notch tension panel made of unidirectional laminates. The crack was assumed to be oriented normal to the applied load direction. Brinson and Yeow [*3*] have measured the fracture stress for these and other configurations. Specifically, their experimental results of relevance here are for $[0]_{8S}$ and $[90]_{8S}$ laminates. For the relatively simple computations performed in this work, it is assumed that there are no debond effects between the piles. Thus, the periodicity condition imposed in the finite-element analysis to reduce computational cost is applicable.

An obstacle that precludes direct comparison between the predictions of the model and the experimental results is that all of the material properties required by the model are not available. For the purpose of the present computations, these properties had to be inferred by making some simplifying assumptions. Specifically, Brinson and Yeow's stress-strain curves for $[0]_{8S}$ and $[90]_{8S}$ lami-

TABLE 1—*Elastic properties used in the simulation of graphite epoxy composite.*

Constituent	Elastic Modulus, E, MPa	Poisson's Ratio, v	Critical Strain Energy Density, MN · m/m³
Fiber	193053	0.3	4.528
Matrix	3413	0.3	0.0855
Interface	3413	0.3	0.0855

nates were assumed to be those of the fiber material and the matrix material, respectively. This assumption was used to obtain the critical strain energy density values given in Table 1.

A further complication was in the interpretation of what Brinson and Yeow have recorded as the fracture stress. It is not clear whether this stress level corresponds to the initiation event or to the onset of unstable crack propagation. This distinction may be particularly important in the $[0]_{8S}$ case in which considerable stable crack growth may precede actual fracture. In the computational results presented here, it is interpreted as the applied stress that, if even slightly increased, results in a large number of elements in the LHR to fail simultaneously. Individual element failure events with increasing applied load are interpreted as indicative of stable damage growth.

For the $[0]_{8S}$ case, computations began by finding the boundary conditions at the LHR boundary using Eq 1. For this case, the roots S_1 and S_2 of the characteristics Eq 3 were found to be

$$S_{1,2} = \mp 1.0857 + 1.9821 \, i$$

The other material constants were found by use of Table 1. The boundary conditions obtained through Eq 1 were then applied to the three-dimensional

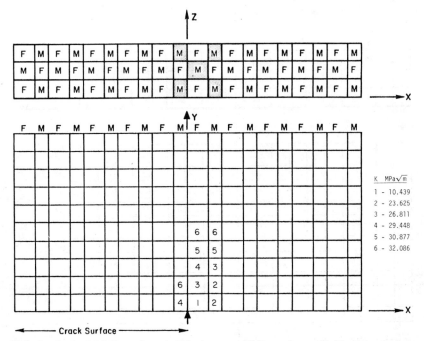

FIG. 6—*Sequential failure of matrix (M) elements with increasing applied load for the 0-deg case.*

TABLE 2—*Comparison of predicted and experimental results for double-edge-notch specimens for $\theta = 0$-deg.*

$\dfrac{2a}{W}$	Predicted Failure Stress, MPa	Experimental Failure Stress, MPa
0.11	435	490.2
0.21	317.8	443.3
0.31	262	274.4
0.41	225.4	263.4
0.51	195.8	211.6

LHR model shown in Fig. 6. The first to fail was a matrix element at Location 1 in Layer 2 at an applied stress intensity factor (K_I) of 10.4 MPa $\sqrt{m}$. Subsequent increases in K_I caused further matrix element failure in the order shown in Fig. 6.

It is clear that the model predicts damage growth parallel to the direction of the fibers and in the matrix elements. This was also observed in the experiments of Ref 3. Stable matrix damage growth continued until the applied stress intensity factor reached 32.4 MPa $\sqrt{m}$. At this stage, 15 new matrix elements failed. This was considered to be an indication of unstable crack propagation.

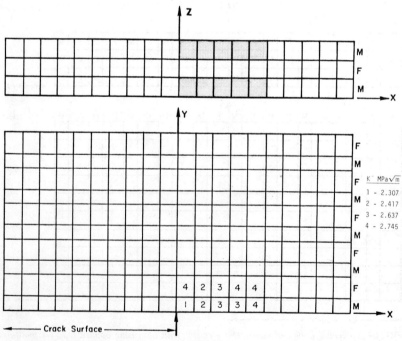

FIG. 7—*Sequential failure of matrix elements with increasing applied load for the 90-deg case.*

TABLE 3—*Comparison of predicted and experimental results for edge-notch specimens for*
θ = *90-deg.*

$\dfrac{2a}{W}$	Predicted Failure Stress, MPa	Experimental Failure Stress, MPa
0.11	37.2	21.4
0.21	26.8	15.8
0.31	22	12.4
0.41	19.3	9.6
0.51	16.5	8.3

Using the value of K_I at the onset of instability, the prediction of fracture stress for different crack lengths in a double-edge-notch specimen of 0.025-m width was made. These are contained in Table 2 along with the experimental measurements of Brinson and Yeow [3].

Results of similar computations performed with the initial crack parallel to the fiber direction and lying entirely in the matrix are shown in Fig. 7. These show that damage growth again occurs in the matrix and parallel to the fiber direction, which is consistent with the general observations made in Ref 3. The predicted and experimentally measured values of failure stress are contained in Table 3.

Concluding Remarks

The computational results presented in this paper, although in seemingly reasonable agreement with some experiments, should not be viewed as conclusive. The assumptions that were necessary to perform the computations, particularly with regard to material properties and to the interpretation of fracture stress, preclude the results of these computations as being indicative of the accuracy of the model. The purpose of the computations was mostly heuristic, and in this respect the three-dimensional LHR model can be considered to have performed well.

References

[1] Kanninen, M. F., Rybicki, E. F., and Griffith, W. I., "Preliminary Development of a Fundamental Analysis Model for Crack Growth in a Fiber Reinforced Composite Material," *Composite Materials: Testing and Design (Fourth Conference), ASTM STP 617,* American Society for Testing and Materials, 1977.
[2] Kanninen, M. F., Rybicki, E. F., and Brinson, H. F., *Composites,* Vol. 8, 1977, pp. 17–22.
[3] Brinson, H. F. and Yeow, Y. T., "An Experimental Study of the Fracture Behavior of Graphite Epoxy Laminates," *Composite Materials: Testing and Design (Fourth Conference), ASTM STP 617,* American Society for Testing and Materials, 1977.
[4] Bazant, Z. P. and Cedolin, L., "Fracture Mechanics of Reinforced Concrete," *Journal of the Engineering Mech. Div., Proceedings,* American Society of Concrete Engineers, Vol. 106, No. EM6, Paper 15917, Dec. 1980.
[5] Sih, G. C. and Liebowitz, H., "Mathematical Theories of Brittle Fracture," *Fracture,* Vol. 2, Academic Press, 1968.

Ronald D. Kriz[1]

Influence of Ply Cracks on Fracture Strength of Graphite/Epoxy Laminates at 76 K

REFERENCE: Kriz, R. D., **"Influence of Ply Cracks on Fracture Strength of Graphite/Epoxy Laminates at 76 K,"** *Effects of Defects in Composite Materials, ASTM STP 836,* American Society for Testing and Materials, 1984, pp. 250–265.

ABSTRACT: Quasi-isotropic laminates ($[0/90/\pm45]_s$ and $[0/\pm45/90]_s$) were fabricated from graphite/epoxy and quasi-statically loaded in tension at 76 K until fracture occurred. Fibers in 0-deg plies carry the largest portion of the tensile load; the weaker 90- and 45-deg plies crack at loads much lower than fracture strength. The effect of ply cracks on fracture of load-bearing 0-deg plies was examined to understand how defects affect laminate strength. A generalized plane-strain finite-element model was used to predict stress gradients in the 0-deg ply near the crack tip. Variations in residual stress caused by changes in temperature and absorbed moisture were included in the analysis. The experiments indicated that absorbed moisture significantly alters the fracture strength and fracture surface of a dehydrated $[0/90/\pm45]_s$ laminate tested at 76 K. The 0-deg plies of dehydrated laminates fractured along several 90-deg ply cracks. When moisture saturated a $[0/90/\pm45]_s$ laminate, a single 90-deg ply crack dominated the fracture of the 0-deg ply and the laminate fracture strength decreased 8%. Analysis of residual stresses indicated that a higher residual stress state existed near the 90-deg ply crack when moisture was absorbed.

KEY WORDS: composite materials, fatigue (materials), fracture mechanics, graphite/epoxy laminates, low temperatures, cracks, residual stress, finite element

When loaded in tension, graphite/epoxy laminates accumulate damage in the form of ply cracks, delaminations, fiber breaks, fiber-matrix debonding, and matrix cracks. Prior to fracture, 0-deg plies carry a larger portion of the tensile load than the weaker 90- and 45-deg plies. Laminate fracture occurs when fibers fracture within the load-bearing 0-deg plies.

Experimental evidence of graphite fiber breaks occurring in 0-deg plies prior to fracture was observed by Kriz [1][2] at room temperature. An enlargement of these fiber breaks is shown in Fig. 1. The fiber breaks observed in Fig. 1 are not influenced by the stress concentration of a 90-deg ply crack. The scatter in

---

[1] Materials research engineer, Fracture and Deformation Division, National Bureau of Standards, Boulder, Colo. 80303.

[2] The italic numbers in brackets refer to the list of references appended to this paper.

FIG. 1—*Graphite fiber breaks adjacent to 90-deg ply crack in a* [0/90/±45]ₛ *graphite/epoxy laminate loaded in tension at room temperature (see Fig. 19 in Ref* 1).

graphite fiber breaks is similar to those modeled by Rosen [2] for unidirectional glass/epoxy. However, Hartwig [3] observed an accumulation of graphite fiber breaks near 90- and 45-deg ply cracks at 76 K. Hence, the presence of large residual stresses near a ply crack at 76 K can influence the fracture of 0-deg plies.

In this study, the influence of residual stress on 0-deg ply fracture was investigated for two quasi-isotropic laminates: $[0/90/\pm45]_s$ and $[0/\pm45/90]_s$. Residual stresses were altered, prior to fracture at 76 K, by dehydrating and saturating laminates with absorbed moisture. A statistical analysis of laminate fracture populations was designed to model the weakening of load-bearing 0-deg plies. A generalized plane-strain finite-element model was used to predict variations in 0-deg ply stresses caused by mechanical and thermal-moisture loads.

Analysis

Stresses within each layer were predicted for a symmetrical laminate loaded in tension. Material response was assumed to be linear and elastic at 76 K. Hence, thermal and moisture loads were superposed with mechanical loads. The applied loads were assumed to be steady and uniform across the laminate. A typical symmetric laminate loaded in tension is shown in Fig. 2, where $L > B \gg T$. Each layer was assumed homogeneous with transversely isotropic elastic properties. Room-temperature and 76-K elastic properties used in the analysis are listed in Table 1. Coefficients of thermal and moisture expansion (listed in Table 1) were modeled as the only contributions to residual stress. Changes in residual stress caused by variations in elastic properties were neglected.

Generalized Plane Strain

Stresses near a ply crack tip ($x = y = 0$, $z/T = 0.75$) were predicted within the x-z plane far from the laminate edges ($y = \pm B$). The constraining out-of-plane strain, ϵ_y, was modeled as the generalized strain normal to the x-z plane

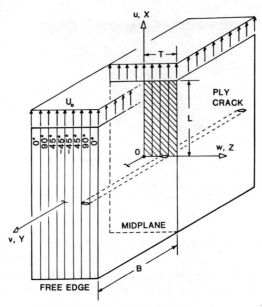

FIG. 2—*Quasi-isotropic laminate dimensions.*

at $y = 0$. The strain in the load direction, ϵ_x, was assumed to be uniform. Using classic laminate-plate theory (see Jones [5]), the strains, ϵ_x and ϵ_y, can be calculated in terms of the known laminate loads, N_x, with $N_y = 0$.

$$\epsilon_x = A_{yy}N_x/(A_{xx}A_{yy} - A_{xy}^2), \quad \epsilon_y = -A_{xy}N_x/(A_{xx}A_{yy} - A_{xy}^2) \qquad (1)$$

where A_{xx}, A_{yy}, and A_{xy} are laminate midplane stiffnesses. Since generalized plane strain was assumed normal to the x-z plane, all six components of stress must be independent of the y-axis. Hence, the equations of equilibrium reduce to

$$\frac{\partial\sigma_x}{\partial x} + \frac{\partial\tau_{xz}}{\partial z} = \frac{\partial\tau_{xy}}{\partial x} + \frac{\partial\tau_{yz}}{\partial z} = \frac{\partial\tau_{xz}}{\partial x} + \frac{\partial\sigma_z}{\partial z} = 0 \qquad (2)$$

The displacement relationships are derived by Talug [6] for the generalized plane-strain problem assuming laminate symmetry

$$\begin{aligned}
u(x,y,z) &= U(x,z) \\
v(x,y,z) &= \epsilon_y y + V(x,z) \\
w(x,y,z) &= W(x,z)
\end{aligned} \qquad (3)$$

where U, V, and W are unknown displacement functions. Talug [6] derived the

KRIZ ON PLY CRACKS 253

TABLE 1—Lamina properties.

	E_1^b	$E_2 = E_3$	$G_{12} = G_{13}$	G_{23}^b	$\nu_{12} = \nu_{13}$
Transversely Isotropic Elastic Properties, GPa	130/143	9.70/18.4	5.39/15.6	3.25/6.18	0.308/0.308
Room temperature[a]/76 K (% increase, [4])	(1.1)	(2.9)	(1.9)	(1.9)	(0)
Thermal Expansion Coefficients, μm/m/K [4]	α_1 −0.128	α_2 8.26			
Moisture Expansion Coefficients, μm/m/% water [1]	β_1 0.0	β_2 3710			

[a] Room-temperature elastic properties from Ref 1.
[b] Subscripts: (1) indicates fiber direction and (23) indicates transverse plane.

elliptic equations of equilibrium in terms of these unknown functions and solved for displacements near ply cracks using finite difference techniques. In this study, thermal and moisture loads were included in the analysis and a finite-element model was used to solve for displacements near ply cracks.

Finite-Element Model

Constant-strain elements with three nodes were chosen. Hence, the required displacement fields (Eq 3) are satisfied for each element by the linear relationships

$$
\begin{aligned}
u &= a_1 + a_2x + a_3z \\
v &= a_4 + a_5x + a_6z + \epsilon_y y \\
z &= a_7 + a_8x + a_9z
\end{aligned}
\tag{4}
$$

where the a_i terms are evaluated in terms of nodal displacements. Elemental stiffness and load matrices for mechanical and thermal-moisture loads were derived by Renieri [7]. A finite-element computer program was written using these elements. Because of laminate symmetry, only the cross-hatched quadrant shown in Fig. 2 was modeled. The grid of elements shown in Fig. 3 was used to model stress concentrations within the 0-deg ply near the 90-deg ply crack. The dimension of the smallest element at the crack tip was less than one graphite-fiber diameter (see Region A of Fig. 3).

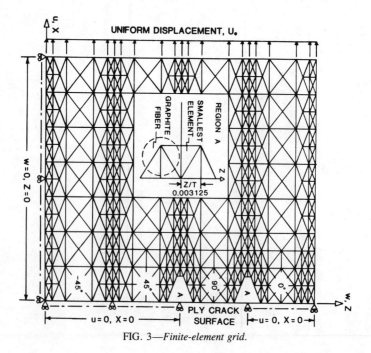

FIG. 3—*Finite-element grid.*

TABLE 2—*Comparison of finite-element and laminate-plate-theory layer stresses.*

Layer	σ_x	σ_y	τ_{xy}
	Stress (MPa), finite element[a]/laminate-plate theory [5]		
0-deg	4.99/5.09	574/577	0.15/0
90-deg	−117/−117	68.9/69.6	0.14/0
45-deg	54.4/55.8	208/212	−94.6/−96.5
−45-deg	54.2/55.8	208/212	94.2/96.5

[a]Finite-element stresses averaged over four elements at each midlayer.

Traction-free boundary conditions along $z = T$ are approximated by prescribing statically equivalent zero nodal forces. Laminate symmetry along $z = 0$ requires displacements $w = 0$. Similarly, displacements $u = 0$ along $x = 0$ are required, except for the traction-free surface $0.5 < z/T < 0.75$ that requires statically equivalent zero-nodal forces. In the x-direction along the surface $x = L$, a uniform axial strain, ϵ_x, is modeled by prescribing a uniform nodal displacement $U_0 = \epsilon_x L$.

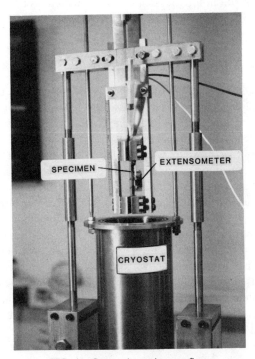

FIG. 4—*Cryogenic tension test fixture.*

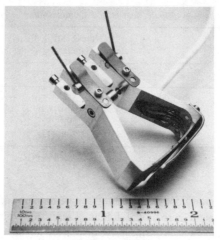

FIG. 5—*Side-mounted extensometer.*

To check the accuracy of the grid shown in Fig. 3, finite-element stresses predicted with no ply crack were compared with thin-laminate-plate theory. This comparison is shown in Table 2 when a grid aspect ratio of $L/T = 6$ is prescribed. Using the grid shown in Fig. 3, finite-element predictions of stress far from the ply crack ($x = 5L$) also agree well with thin-laminate-plate theory.

Procedure

A total of 73 $[0/\pm45/90]_s$, 73 $[0/90/\pm45]_s$, and 70 $[0_8]$ specimens were fabricated from the same batch of graphite/epoxy reported in Ref *1*. All specimens were 10.2-cm long with uniform width. Quasi-isotropic laminates were 1.27-cm wide and unidirectional specimens were 0.64-cm wide. Half of the specimens of each type were exposed to 95% relative humidity at 343 K until no additional absorbed moisture weight gain could be measured (wet or saturated condition). The remaining specimens were dried in an oven at 338 K until no additional moisture weight loss could be measured (dry or dehydrated condition). The edges of four $[0/\pm45/90]_s$ and four $[0/90/\pm45]_s$ were polished and replicated as outlined in Ref *8* prior to wet and dry conditioning. Thirty-five wet and 35 dry quasi-isotropic laminates (gage length 7.62 cm) were quasi-statically loaded in tension to fracture using a hydraulic load controlled system. The cryogenic tension load fixture is shown in Fig. 4. A special extensometer (shown in Fig. 5) was used to measure strain over a 1.27-cm gage length with $\pm0.05\%$ accuracy. Glass/epoxy tabs, 1.27-cm long, were used to grip the quasi-isotropic specimens with a fine-mesh stainless-steel interface. The stainless-steel mesh allowed thermal contraction mismatch between graphite/epoxy specimens and glass/epoxy tabs without causing unnecessary stress concentrations. Glass/epoxy

TABLE 3—*Weibull and normal strength distributions at 76 K.*

Laminate Configuration	Weibull Strengths, MPa					Normal Strengths, MPa		
	α	β	Mean	Standard Deviation	%[a]	Mean	Standard Deviation	%[a]
[0$_8$] dry	5.07	1280	1180	267	80	1180	241	60
[0$_8$] wet	3.83	1150	1040	303	92	1040	267	80
[0/90/±45]$_s$ dry	17.6	475	460	32.3	92	461	29.1	60
[0/90/±45]$_s$ wet	19.0	425	425	27.7	93	425	25.2	85
[0/±45/90]$_s$ dry	12.6	376	361	34.8	90	361	32.1	96
[0/±45/90]$_s$ wet	16.3	387	375	28.3	50	375	25.3	70

[a]Percent confidence from a chi-square test [10].

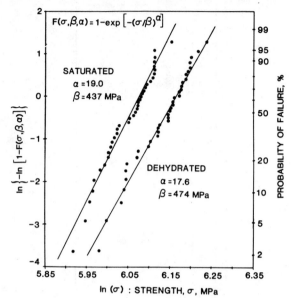

FIG. 6—*Weibull plot of wet and dry* $[0/90/\pm45]_s$ *strengths.*

tabs were bonded to all $[0_8]$ specimens with epoxy adhesive. No debonding was observed at 76 K.

All specimens with polished edges were loaded in increments of 445 N, and replicas were taken of edge damage at room temperature after each load increment. Hence, the load increment required to initiate ply cracking was recorded. Thermal cycling after each load increment resulted in no new damage.

Results and Discussion

All specimens failed at random locations between grips. There were no apparent stress concentrations near tabs that influenced fracture strength. Strengths reported in Table 3 compare well with results reported in Ref 4. Hence, strengths recorded in Table 3 are minimally influenced by test procedure. Statistical-distribution functions used to represent inherent scatter should not be chosen a priori [9]. Here we used normal and Weibull distribution functions and tested for "goodness of fit" using a chi-square test [10]. All $[0_8]$, $[0/90/\pm45]_s$ and $[0/\pm45/90]_s$ specimens fit the Weibull distribution best with 80% confidence or better. Weibull distributions imply a weak-link effect that influences fracture strength.

The largest difference between wet and dry strengths was observed for $[0/90/\pm45]_s$ laminates. Weibull strength distributions for wet and dry $[0/90/\pm45]_s$ laminates are shown in Fig. 6. Comparison of Weibull shape parameters $\alpha_{WET} > \alpha_{DRY}$ indicates a more dominant weak-link effect when moisture is ab-

DRY

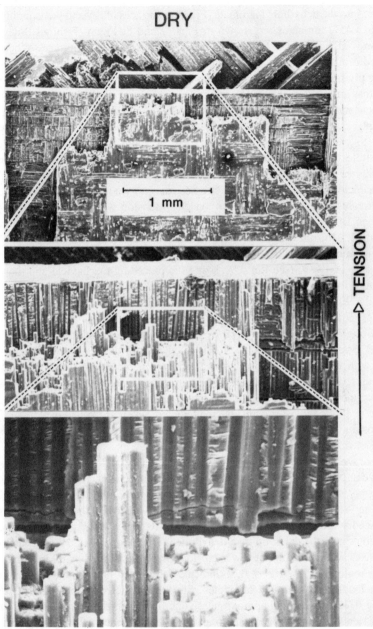

TENSION

FIG. 7—*Fracture surface of a dry [0/90/ ±45]ₛ laminate.*

sorbed into the dehydrated laminate. Fracture surfaces of dry and wet conditioned $[0/90/\pm45]_s$ laminates are shown in Figs. 7 and 8. Figure 7 shows 0-deg ply fractures along several 90-deg ply cracks of a dehydrated laminate. When absorbed moisture saturates the laminate, a single 90-deg ply crack dominates the 0-deg ply fracture shown in Fig. 8 and the Weibull mean strength decreases 8%.

Stress distributions near 90-deg ply cracks were predicted with the finite-element model. The mechanical plate load, N_x, required to initiate the first 90-deg ply crack was observed at $N_x = 273$ kN/m. Relationships in Eq 1 were used to calculate strains, $\epsilon_x = 4010$ μm/m and $\epsilon_y = -964$ μm/m, which constitute the mechanical load used by the finite-element model. Stresses in dehydrated laminates loaded in tension were calculated by superposition of the mechanical stresses with residual stresses caused by a temperature change at 76 K. Stresses in saturated laminates loaded in tension were calculated by superposition of additional residual stresses caused by swelling when moisture is absorbed. Moisture weight gain was measured at 1.2% in the saturated state and a temperature change of -318 K was chosen, with the idealization of a stress-free state at 494 K (see Ref 11). Using the preceding mechanical and thermal-moisture loads together with the elastic properties defined in Table 1, stresses near the 90-deg ply crack in a $[0/90/\pm45]_s$ laminate were calculated and plotted in Figs. 9 through 11.

Through-thickness variation of σ_x in the 0-deg ply along $x = 0$ is shown in Fig. 9. Stress in the load x direction is calculated for three cases: (1) mechanical load with no residual stresses, σ_x^{MECH}; (2) mechanical load including residual thermal load, σ_x^{DRY} (dehydrated condition); and (3) mechanical load including residual thermal-moisture load, σ_x^{WET} (saturated condition). At position $z/T = 0.753$ and $x = 0$, the largest σ_x stresses are predicted, where $\sigma_x^{MECH} < \sigma_x^{WET} < \sigma_x^{DRY}$. At position $z/T = 0.766$ and $x = 0$, the inequality of σ_x stresses is reversed, $\sigma_x^{DRY} < \sigma_x^{WET} < \sigma_x^{MECH}$.

Here we assume laminate fracture is dominated by 0-deg ply fracture because the 0-deg plies carry the largest portion of the tensile load. Hence, the predicted 0-deg ply stresses, σ_x^{WET} and σ_x^{DRY}, were compared with the mean wet and dry fracture strengths ($X_{WET} = 425$ MPa, $X_{DRY} = 460$ MPa). In Fig. 9, stresses averaged within the region $0.761 < z/T < 1.0$ for saturated (WET) laminates are larger near the crack tip than dehydrated (DRY) laminates. Hence, saturated laminates would fracture at lower tension loads. Conversely, stresses predicted in the region $0.75 < z/T < 0.76$ indicate dehydrated laminates will fracture at lower loads than saturated laminates. Stresses predicted in the region $0.75 < z/T < 0.761$ are located approximately one fiber diameter from the 0/90 interface. Hence, the heterogeneous graphite/epoxy structure of the 0-deg ply can not be modeled as a continuum within the region $0.75 < z/T < 0.76$. Stresses predicted in this region were ignored.

Delaminations were observed only at the 0/90 interface near the fracture surface. However, delaminations at the 0/90 interface were not observed on replicas taken prior to fracture. Since delaminations occurring at the 90-deg ply

WET

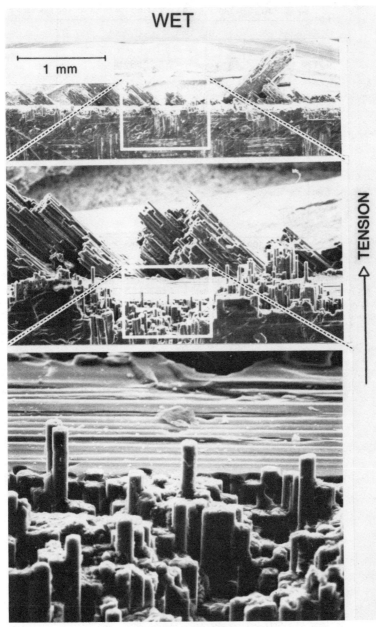

TENSION

FIG. 8—*Fracture surface of a wet [0/90/±45]ₛ laminate.*

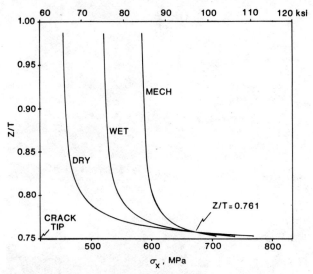

FIG. 9—*Variation of* σ_x *through the 0-deg ply thickness above the 90-deg ply crack tip. Load cases: (1) mechanical load* $N_x = 273$ *kN/m with no residual stress (MECH), (2) mechanical load including residual thermal load* $(-318$ *K) (DRY), and (3) mechanical load including residual thermal and moisture load* $(+1.2\%$ *water) (WET).*

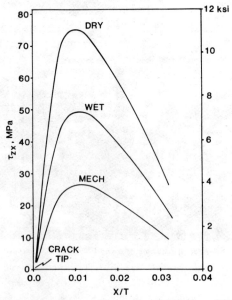

FIG. 10—*Variation of* τ_{xz} *with x along surface* $z/T = 0.766$ *above 0/90 interface. Load cases are the same as shown in Fig. 9.*

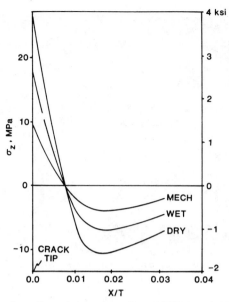

FIG. 11—*Variation of* σ_z *with* x, *along surface* $z/T = 0.766$ *above 0/90 interface. Load cases are the same as in Fig. 9.*

crack tip will blunt the stress concentration, fiber breaks within the 0-deg ply will not occur near the crack tip. In Figs. 7 and 8, we observed broken 0-deg ply fibers near 90-deg ply crack tips. Although delaminations did not influence 0-deg ply fracture, interlaminar stresses were plotted near the 0/90 interface in Figs. 10 and 11. When moisture is absorbed, lower interfacial shear stress, τ_{xz}, is predicted near the crack tip (see Fig. 10). Similarly, stress normal to the 0/90 interface is lower when moisture is absorbed (see Fig. 11). Hence, stresses leading to delaminations are decreased due to absorbed moisture.

Laminates with a stacking sequence of $[0/\pm45/90]_s$ were less affected by absorbed moisture at 76 K than $[0/90/\pm45]_s$ laminates. Absorbed moisture increases $[0/\pm45/90]_s$ laminated strength by 3.8% at 76 K. Fracture surfaces of dehydrated and saturated $[0/\pm45/90]_s$ laminates were similar; 0-deg ply fiber fractures occurred randomly and stepwise along several 45-deg ply cracks as shown in Fig. 12.

Strengths for unidirectional $[0_8]$ specimens, shown in Table 3, scattered more than the strengths of quasi-isotropic laminates. Absorbed moisture decreases strength 12% at 76 K. When unidirectional graphite/epoxy is bonded to 90- and 45-deg plies, a constraining effect on strength was observed by Stinchcomb [12]. Strengths reported in Table 3 for $[0,90,\pm45]_s$ specimens are reduced to equivalent cross-sectional areas of 0-deg plies and compared with $[0_8]$ strengths. This comparison is justified when the 0-deg ply fracture is assumed to dominate laminate fracture strength. Unconstrained $[0_8]$ mean strength is 23% lower than

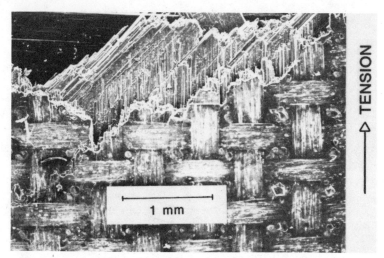

FIG. 12—*Fracture surfaces of a wet* $[0/\pm45/90]_s$ *laminate.*

the fracture strength of a 0-deg ply constrained by 90- and 45-deg plies. Detailed discussions on constraint are given in Refs *6* and *12*.

Conclusions

Predicted dehydrated and saturated residual stresses within 0-deg plies near ply cracks account for the observed differences in $[0/90/\pm45]_s$ fracture strengths, assuming 0-deg ply fracture dominates laminate fracture at 76 K. Absorbed moisture swells 90- and 45-deg plies and changes the residual thermal 0-deg ply stresses. When dehydrated laminates are saturated and loaded in tension at 76 K, larger residual stress exists near the crack tip. Hence, saturated laminates fracture at lower tension loads. Fractographic observations confirm this prediction, where a single 90-deg ply crack is observed to dominate the fracture of a 0-deg ply. This dominance of a 90-deg ply crack on the 0-deg ply fracture is recovered statistically, where the Weibull shape parameter, α, indicates a stronger weak-link effect when $[0/90/\pm45]_s$ laminates are saturated. Laminate strength is increased 8% when dehydrated.

Laminates with the stacking sequence $[0/\pm45/90]_s$ are less affected by absorbed moisture at 76 K than the $[0/90/\pm45]_s$ laminates.

Unconstrained $[0_8]$ unidirectional strengths are lower with more scatter when compared with constrained 0-deg ply strengths. Strength of unidirectional graphite/epoxy at 76 K was improved when constrained by crossplies.

Acknowledgments

This work was sponsored by the National Research Council-National Bureau of Standards Postdoctoral Research Program and the U.S. Department of Energy, Office of Fusion Energy.

References

[1] Kriz, R. D., "Effects of Moisture, Residual Thermal Cure Stresses, and Mechanical Load on Damage Development in Quasi-isotropic Laminates," Ph.D. thesis, Department of Engineering Science and Mechanics, College of Engineering, Virginia Polytechnic Institute and State University, Blacksburg, Va., Dec. 1979.

[2] Rosen, B. W. in *Fiber Composite Materials*, American Society for Metals, Metals Park, Ohio, 1965, pp. 37–75.

[3] Hartwig, G. in *Advances in Cryogenic Engineering—Materials*, Vol. 28, R. P. Reed and A. F. Clark, Eds. Plenum Press, New York, 1982, pp. 179–189.

[4] Haskins, J. F. and Holmes, R. D., "Advanced Composite Design Data for Spacecraft Structural Applications," Technical Report AFML TR-79-4208, Air Force Materials Laboratory, Wright-Patterson Air Force Base, Ohio, Oct. 1979.

[5] Jones, R. M., *Mechanics of Composite Materials*, McGraw-Hill, New York, 1975, pp. 147–152.

[6] Talug, A., "Analysis of Stress Fields in Composite Laminates with Interior Cracks," PhD thesis, Department of Engineering Science and Mechanics, College of Engineering, Virginia Polytechnic Institute and State University, Blacksburg, Va., Aug. 1978.

[7] Renieri, G. D., "Nonlinear Analysis of Laminated Fibrous Composites," PhD thesis, Department of Engineering Science and Mechanics, College of Engineering, Virginia Polytechnic Institute and State University, Blacksburg, Va., June 1976.

[8] Stalnaker, D. O. and Stinchcomb, W. W. in *Composite Materials: Testing and Design (Fifth Conference)*, ASTM STP 674, American Society for Testing and Materials, Philadelphia, 1979, pp. 620–641.

[9] Tracy, P. G., Rich, T. P., Bowser, R., and Tramontozzi, L. R., *International Journal of Fracture*, Vol. 18, No. 4, April 1982, pp. 253–277.

[10] Park, W. J., "Basic Concepts of Statistics and Their Applications in Composite Materials," Technical Report AFML-TR-79-4070, Air Force Materials Laboratory, Wright-Patterson Air Force Base, Ohio, June 1979.

[11] Pagano, N. J. and Hahn, H. T. in *Composite Materials: Testing and Design (Fourth Conference)*, ASTM STP 617, American Society for Testing and Materials, Philadelphia, 1977, pp. 317–329.

[12] Stinchcomb, W. W., Reifsnider, K. L., Yeung, P., and Masters, J. in *Fatigue of Fiberous Composite Materials*, ASTM STP 723, American Society for Testing and Materials, Philadelphia, 1979. pp. 320–333.

Index

161
166

190 localisation
207 complications
    loading
227 ?
236

233